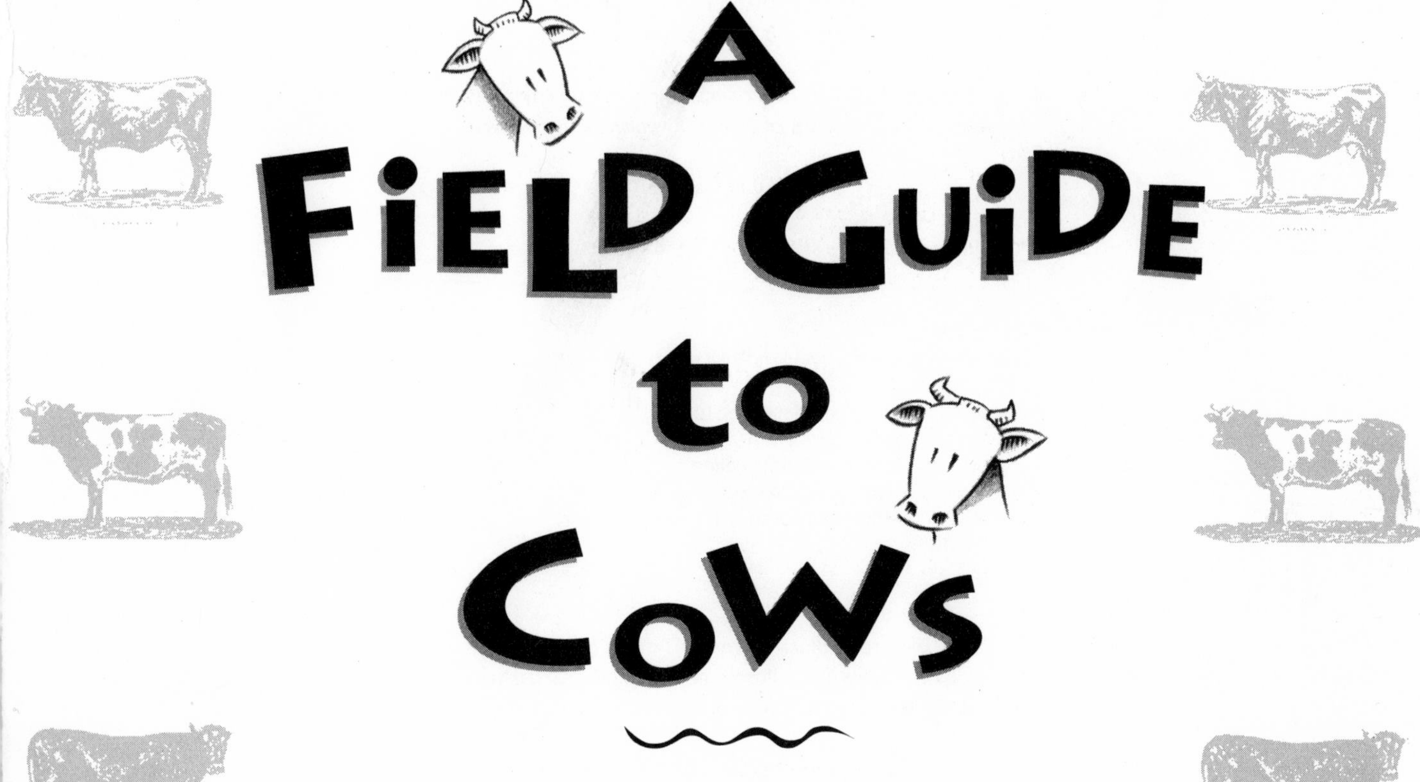

BY JOHN PUKITE

FALCON™
HELENA, MONTANA

Copyright © 1996 by Falcon Press Publishing Co., Inc.
Helena and Billings, Montana

All rights reserved, including the right to reproduce this book or any parts thereof in any form, except for inclusion of brief quotations in a review.

Printed in Canada.

Library of Congress Cataloging-in-Publication Data

Pukite, John, 1964-
A field guide to cows / by John Pukite.
p. cm.
Includes bibliographical references (p.):
ISBN 1-56044-424-X (pbk.)
1. Cattle--United States--Identification. 2. Cattle--Canada--Identification. 3. Cattle--United States--Pictorial works. 4. Cattle--Canada--Pictorial works. I. Title.
SF198.P84 1996
636.2--dc20 96-12469
CIP

Cow breed illustrations by Todd Telander.
Cartoons and Cow Fact illustrations by Peter Grosshauser.
Front cover photo: University of Wisconsin, Department of Agricultural Journalism.

The poem "The Cow" is from VERSES FROM 1929 ON by Ogden Nash.
Copyright 1931 by Ogden Nash. Copyright © Renewed. First appeared in THE NEW YORKER.
By permission Little, Brown and Company.

Excerpted lines from Rudyard Kipling's poem "Alnashar and the Oxen" from the book *Rudyard Kipling's Verse Definitive Edition* printed by permission of Bantam Doubleday.

The Far Side cartoon by Gary Larson is reprinted by permission of Chronicle Features, San Francisco, CA. All rights reserved.

For extra copies of this book and information about other Falcon books, write Falcon Press, P.O. Box 1718, Helena, MT 59624; or call 1-800-582-2665.

♻ Text pages printed on recycled paper.

Contents

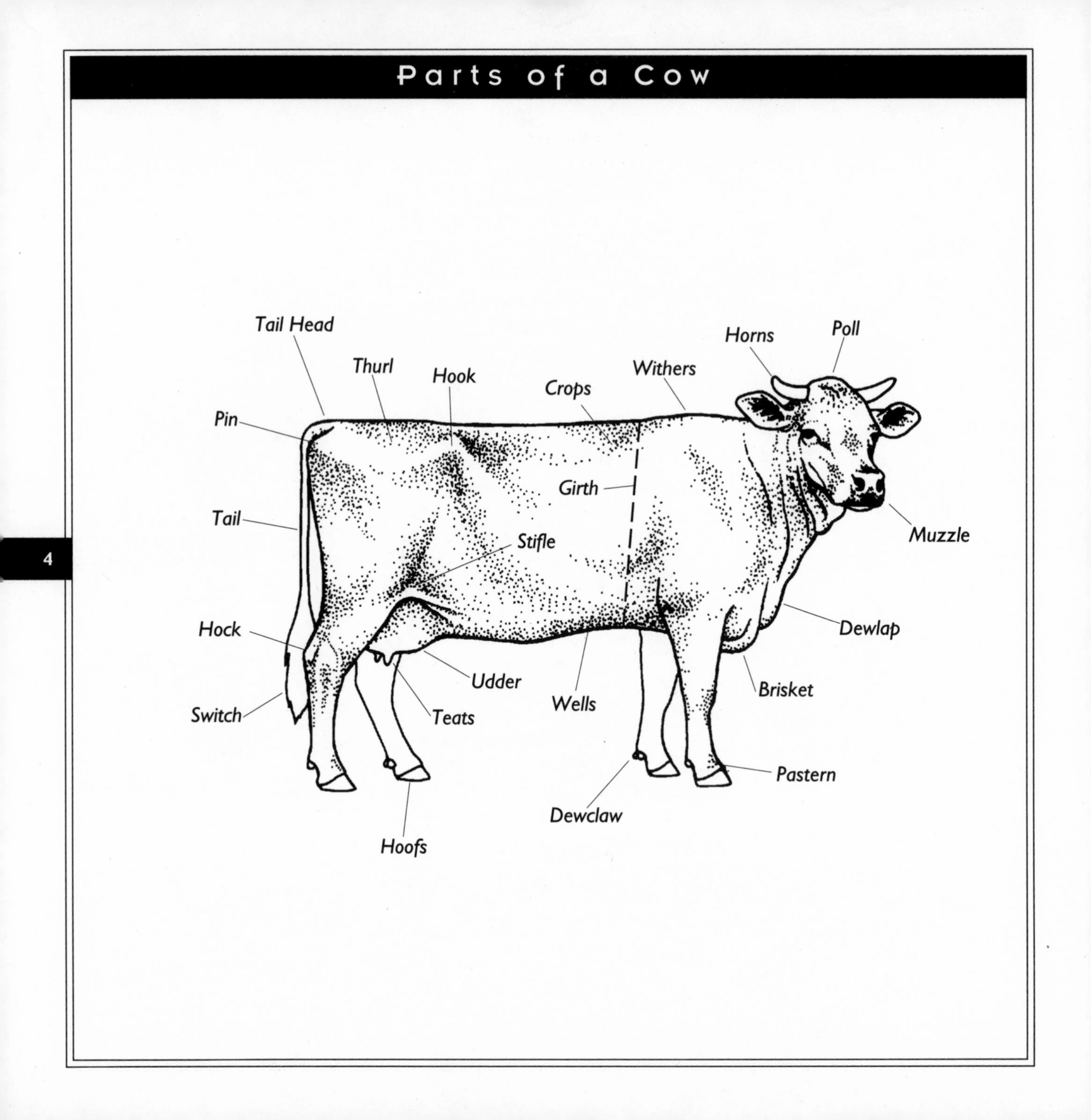
Parts of a Cow
Tail Head
Thurl
Hook
Crops
Withers
Horns
Poll
Pin
Girth
Tail
Stifle
Muzzle
Hock
Dewlap
Udder
Wells
Brisket
Switch
Teats
Pastern
Dewclaw
Hoofs
4

WHY WATCH COWS?

At first the idea might seem foolish, but with about 100 million cattle in the United States, the lowly cow is one of the most numerous and popular animals around. As such, cows are easy to find, readily identifiable, and enjoyable to view and hear moo. Cow watchers can find a rich assortment of breeds that originated from all parts of the world, and can learn how the breeds have adapted to different regions. As you go along, you will discover the thrill of spotting a rare breed or completing a checklist. If you are already an amateur biologist, cow watching can add a fun dimension to your notes; include, for example, the type of cow that cattle egret happens to be sitting on. Whether you are a farmer, biologist, or vacationing family traveling through the countryside, you should try your hand at cow watching—the perfect hobby for anyone who wants to know more about bovines.

HOW TO IDENTIFY COWS

So you've spotted a herd of cows contentedly chewing their cud in a pasture. You ask yourself, "What kind are they?" If you're an experienced cow spotter, the answer might come right away; if you're not, take a good look at the cows and also at their environment. Both are clues to the cows' breed.

Here are some observations you may want to make. Look at the coloration of a cow or look at the color of the whole herd if they are all similar. Determine the color category the cow fits into. Check the cow's body shape, or conformation—is it wedged like dairy cattle, rectangular like beef cattle, or somewhere in between? How big is the cow? Look for other features, such as the type of horns or the lack of horns (a polled cow). Does it have a hump? Does it have long hair? Use these clues and the "Key to the Cows" in this guide to lead you to the correct breed.

The answers to these questions should lead to a positive ID, but you might want to verify the breed with some followup questions about the cow's environment. For example, what is the local climate? Some breeds, like the Brahman, like hot weather. Other breeds do best in cold climates. What kind of land are the cows grazing on? If it is wide open

Beef (rectangle outline)

Dairy (wedge-shaped outline)

HOW TO IDENTIFY COWS

rangeland, more than likely they are beef cattle. A fenced-in pasture probably means they are a dairy or dual-purpose breed. Don't forget to look for easy clues like signs advertising the breed a farmer raises. Some might call that cheating; I call it using all available resources.

If you decide on the cow's breed, and everything seems to match up—excellent; but don't despair if something seems out of kilter. The cow might be a cross of a couple or more different breeds. Farmers and ranchers do this quite a bit to improve or change the makeup of their herd. Take a look at the crossbreeding chart to get an idea of how many combinations are possible. The cow could also be quite simply an unimproved or scrub cow. Scrub cattle are any odd mixture of cattle of unknown lineage or breeding history—sort of the polite way of saying "mutt." Some traits might show through enough to give you an idea of the cow's origin. For example, big ears or loose, hanging skin often indicate some parentage from the Brahman.

Cow Fact:

A cow has four stomachs. In descending order they are:

1) *Rumen or paunch – holds thirty gallons;*
2) *Reticulum or honeycomb;*
3) *Omasum or manyplies;*
4) *Abomasum – considered the only true stomach.*

CLOTHING AND EQUIPMENT

Like in bird watching, not much gear is needed to spot cows. In fact, all you really need are your eyes. Of course, this is the bare minimum, and sooner or later you'll probably want a decent pair of binoculars to identify that far-off wandering cow. I would recommend a pair of at least 6x35 power binoculars. The first number is magnification power: in this case the cow will appear six times larger than with the naked eye. The second number refers to the diameter of the front lenses—bigger numbers give you more light-gathering power, which is important for early morning or late evening viewing. When purchasing binoculars, go for the highest quality you can afford, but don't necessarily go for greater magnification. Often a high-quality lens at a lower magnification will provide a better image than a less expensive but more powerful lens.

Other possible equipment is limited to a notebook and pencil for recording notes or drawing spotting patterns. Some might think about purchasing a commercial cow caller—one of those devices that moo when you turn it over. I have devised a much simpler and more effective call that achieves better results and more volume. (A limited edition version of this call has been included with this book.)

As for clothing I prefer blue jeans and a seed cap (Mallard Seeds). Since I live in the Midwest this is a perfect disguise for the cows who think I must be their farmer. Cow fanciers in different parts of the country might want to try farm overalls or a cowboy hat.

And don't forget your footwear. The options here are plentiful: old shoes, galoshes, Wellingtons, or acid-resistant boots—horse-hide leather is the best. The large amount of manure in a cow pasture makes the soil's composition fairly acidic, and that has a tendency to destroy shoes fairly quickly. For this reason, unless you're on horseback, wearing cowboy boots in dairy country is probably not the best idea.

VIEWING TIPS

Straightforward as watching cows might be, a few helpful hints might make spotting cows a little easier and more enjoyable. Cows are part of the family Bovidae (refer to the taxonomic chart of cows). This means they share some of the same traits and behaviors as their wilder relatives like the Bison.

Cows have almost total 360-degree panoramic vision, making them tough to sneak up on. So don't even try. It's much better to simply move slowly up to them without making any sudden movements or loud noises. Usually you can get quite close, which might mean up to the fence they happen to be grazing alongside. If stealth isn't up your alley, your car can also be a fine place to observe the ruminations of the beasts. Some people like to watch cows from horse-back, which could be helpful out on the range, but most cow spotting can be done from the comfort of your own car or truck. With a pair of binoculars the view can be just as good as on foot and you won't have to worry about chasing them to the far side of the pasture.

In the morning and evening you might see them heading out and back from the fields near their barn. At midday, take a look to see if they are relaxing under shady trees, cooling themselves and chewing their cud. Keep your eyes open for trees with parts worn smooth by cows scratching themselves. In spring and fall out in the western states, you might catch them during their annual migration, which brings them from higher summer grazing grounds to lower winter grounds. You can also find cows by following their tracks or by spotting cow pies, which are fairly large, somewhat flat, and made up of lots of partially digested forage matter.

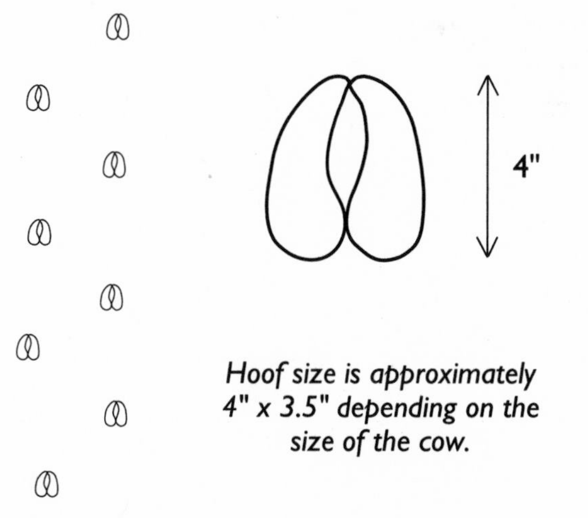

Hoof size is approximately 4" x 3.5" depending on the size of the cow.

CAUTIONS

When you're dealing with any animal that can weigh over a ton and has horns, you should take a few precautions. Some of these depend on the type of cow. A dairy cow that lives around people all the time will be much more friendly than a beef cow that's been out on the range for a month. Stay alert during calving season because the mother cows are more protective of their calves and might defend them against a stranger. Keep away from bulls! Males of some breeds can be especially ornery. If a cow has horns, watch out for a swing of the head. The cow might be just trying to scratch when you get in the way of a horn tip. Unless a cow's tail is tied to its leg, take heed of being kicked.

Watch your step—a cow pie can cause an unforeseen spill at the worst moment, like right when a bull starts chasing you. Don't wear the color red. Even though they say cows don't see colors the same way we do, there's no use taking the chance of antagonizing a cow. (Then again, barns, bandannas, and a few cattle breeds are red, and that doesn't seem to bother the cows too much.)

Other features to be aware of include electric fences, barbed-wire fences, cattle guards, and those cow dogs that don't take kindly to people poking about.

COW ETIQUETTE, OR THE DOS AND DON'TS OF RESPONSIBLE COW WATCHING

While there are no hard and fast rules to watching cows, there are a few general guidelines any responsible bovine enthusiast should follow.

Cows are often on private land, which means you should respect the landowner's property. You can probably get permission to look around a bit from the farmer (find the farmer at the homestead or a nearby cafe). If you get permission, make sure you leave the gates as

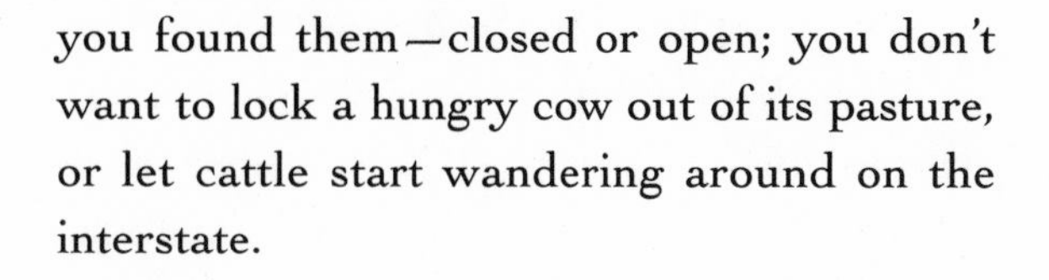

you found them—closed or open; you don't want to lock a hungry cow out of its pasture, or let cattle start wandering around on the interstate.

DON'T moo excessively at cows with calves. Mooing, or lowing as it is also called, can signal distress or act as a "feed me" cry from the calf. Otherwise, mooing generally does no harm.

DON'T chase cows. They just might turn and chase you.

DON'T cow tip. Besides the obvious dangers of this sport (the cow might fall on you, it might wake up and chase you, or gore you with its horns), cow tipping can be unhealthy for the cow if you really did manage to knock it over. A warning to those foolish enough to try such an act: Remember, cows often weigh in excess of 2,000 pounds, they have a low center of gravity, and they never really sleep standing up—for that they lie down.

How to Use This Guide

A Field Guide to Cows brings together for the first time everything needed to identify cows grazing on North American farms and ranches. In addition to the thorough description of each breed, the guide gives detailed information on the breed's purpose and characteristics, as well as the breed's history. The guide organizes all this information into simple categories based on the most easily identifiable cow traits, such as coloration pattern or body shape. Each breed description provides the following information.

Synonyms: Alternative names for the breed, both current and obsolete.

Distribution: The approximate number of registered cattle in North America.

- ***Critical***—Less than 200 registered cattle.
- ***Rare***—Between 200 and 1,000 registered cattle.
- ***Sparse***—Between 1,000 and 2,500 registered cattle.
- ***Common***—Numerous and often seen; above 2,500 registrations.
- ***Ubiquitous***—Widespread and always seen, above 50,000 registrations.
- ***Limited***—Not common in the United States, but numerous elsewhere in the world.

Description: Specific identifying traits of each cow breed in terms of color, body conformation (shape), horns, and size, plus any other unique characteristics that point to differences in related breeds.

Purpose: The main uses of the cattle breed, both present and former, with the categories of beef, dairy, and draft. Cows labeled as dual or triple purpose combine these categories, meaning they have multiple uses on a farm. This section may also list the traditional and popular cheeses coming from the cow's native land.

Origin: The location of the breed's first husbandry, plus known information on the breed's development. Includes dates of importation to America and of the formation of the cow's breed society.

Drawings: Sketches of an adult cow and bull that have arrows pointing out the breed's unique characteristics. Also presents the average height and weight of mature animals.

Key to the Cows *Bovidae*

Color	Field Marks and Purpose	Page
Black & White	**1. Spotted**	
	Holstein—Big black spots. Dairy.	16
	Black Baldy—White face and black body. Beef.	18
	2. Belted	
	Dutch Belted—White belt on black. Dairy.	20
Black	**1. Long Horns**	
	Highland—Shaggy with long horns. Beef and Draft.	22
	2. Polled	
	Aberdeen-Angus—Stocky, sleek body. Beef.	24
	Galloway—Curly hair. Beef and Dairy.	26
	3. Medium-Length Horns	
	Welsh Black—Long body on short legs. Beef and Dairy.	28
	Dexter—The smallest cow. Dairy, Beef, and Draft.	30
	Canadienne—Milk cow conformation. Dairy.	32
Red & White	**1. Long Horns**	
	Corriente—Multicolored, small, and fit. Rodeo.	34
	Texas Longhorn—Lyre-like horns. Beef.	36
	Watusi—The biggest horns. Horns.	38
	2. Beef Conformation	
	Hereford—White, wide face and deep red body. Beef.	40
	Simmental—White face and yellowish red body. Beef and Dairy.	42
	Pinzgauer—Brown-sided. Beef and Dairy.	44
	Normande—Tricolored with spectacles. Beef and Dairy.	46
	MRY—Like a beefy red-and-white Holstein. Beef and Dairy.	48
	Maine-Anjou—White triangle on forehead. Beef and Dairy.	50
	Shorthorn—Roan and beefy. Beef.	52

KEY TO THE COWS

COLOR	FIELD MARKS AND PURPOSE	PAGE
RED & WHITE (CONTINUED)	**3. Dairy Conformation**	
	Milking Shorthorn—Roan and milky. Dairy.	54
	Norwegian Red—Mostly red. Beef and Dairy.	56
	Ayrshire—Lots of smaller spots. Dairy.	58
	Guernsey—Fawn spots and golden tears. Dairy.	60
BROWN	**1. Dairy Conformation**	
	Jersey—Small and doelike. Dairy.	62
	Brown Swiss—Large body and fuzzy ears. Dairy.	64
	2. Beef Conformation	
	Gelbvieh—Yellow and muscular. Beef and Dairy.	66
	Blonde d'Aquitaine—Big blonde. Beef.	68
	Limousin—Long, sleek body. Beef.	70
RED	**1. Big Reds**	
	Tarentaise—Black nose. Beef and Dairy.	72
	Salers—Triangle-shaped head. Beef and Dairy.	74
	South Devon—Sandy red color. Beef and Dairy.	76
	Devon—Yellow-orange skin. Beef, Dairy, and Draft.	78
	Sussex—Big shoulders and tough looking. Beef.	80
	Red Poll—Always polled. Beef and Dairy.	82
	Lincoln Red—Like a solid red Shorthorn. Beef and Dairy.	84
	Danish Red—Dark skin. Dairy and Beef.	86
WHITE	**1. Stocky Shape**	
	White Park—Long horns; small, shaggy body. Beef and Wildness.	88
	British White—Polled and gentle. Beef and Dairy.	90
	Murray Grey—Polled and classically shaped. Beef.	92
	Charolais—Big with a pink nose. Beef.	94

KEY TO THE COWS

COLOR	FIELD MARKS AND PURPOSE	PAGE
WHITE (CONTINUED)	**2. Big and Tall Podolians**	
	Piedmont—Double-muscled butt. Beef.	96
	Chianina—The biggest cow. Beef.	98
	Romagnola—Rough looking with longish horns. Beef.	100
	Marchigiana—Straight topline and small horns. Beef.	102
ZEBU (HUMPED COWS FROM ASIA AND AFRICA)	**1. Big Hump (*Bos indicus*)**	
	Brahman—Slanting rump; loose, hanging skin. Beef.	104
	2. Bos indicus* x *Bos taurus	
	Brangus—Black and baggy. Beef.	106
	Braford—White face with red spectacles. Beef.	108
	Beefmaster—Red or red-and-white; big and beefy. Beef.	110
	Santa Gertrudis—Solid red. Beef.	112
	Barzona—Red; rectangular face with oval horns. Beef.	114
BISON (WILD NORTH AMERICAN COWS)	**1. Big Hump (*Bison bison*)**	
	Bison—Large, shaggy head; narrow waist. Beef.	116
	2. *Bison* x *Bos taurus*	
	Beefalo—Variable color and conformation. Beef.	118

?!

- *The world's biggest Bison (statue) is in Jamestown, North Dakota.*
- *The world's biggest Holstein (statue) is in New Salem, North Dakota.*
- *The world's biggest talking cow (Holstein statue) and a 17-ton replica of the world's biggest cheese are in Neillsville, Wisconsin.*

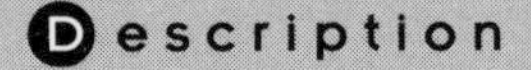

Description

THE BLACK-AND-WHITE, SPOTTED COW.

The spots usually cover about 50 percent of the cow's body, but actual coverage may range from all to none (i.e., from an all-white cow to an all-black cow). The spotting usually makes identification easy, but other traits to look for include a broad face; a wide muzzle; a wedgy dairy shape; a tall, stately stance; and a clean-cut appearance. Also, the Holstein is almost always polled. A few recessively red-spotted cows might be mixed in an otherwise all-black-and-white herd.

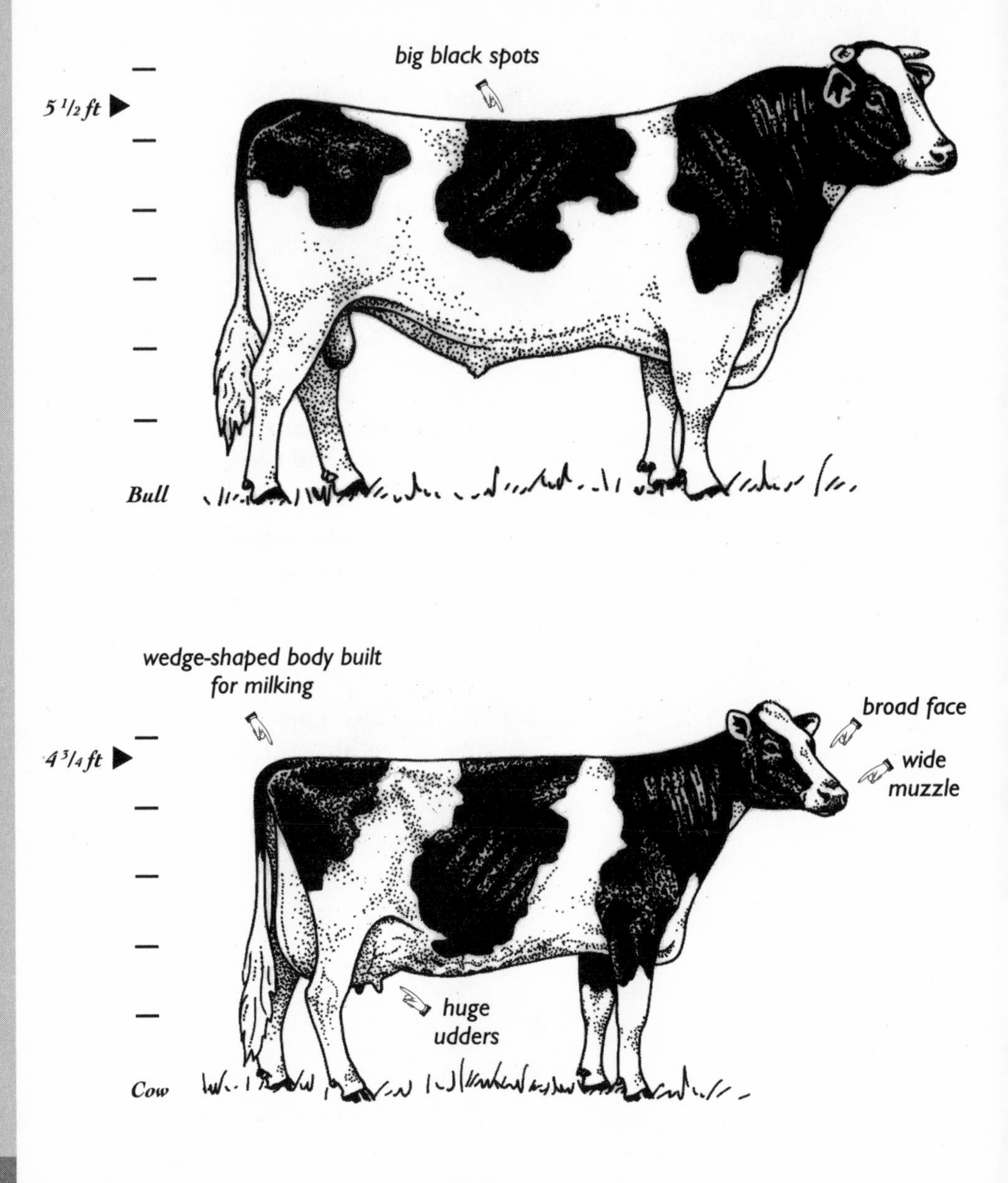

spotted black & white

Holstein

- Synonyms: ***Holstein-Friesian, American Friesian, Dutch Black Pied***
- Distribution: ***Ubiquitous***
- Registrations: ***over 500,000***
- Bull's Average Weight: ***2,200 pounds***
- Cow's Average Weight: ***1,500 pounds***

Purpose

Dairy

The Holstein produces the most milk of all the breeds. Farmers raise the Holstein solely for dairy production, though it ends up as a de facto beef cow at the end of its short, 6-year working lifespan. The huge bulls once worked as oxen; one famous ox supposedly weighed in at 4,365 pounds and stood 17 ½ hands. As for the cow's tremendous milk yields, it achieves them partly through its genetic makeup and partly through intensive management: farmers provide concentrated feed, lots of veterinary care, and a nice, environmentally controlled barn. The production record, made by the cow Raim Mark Jinx, was over an amazing 100,000 glasses of milk.

Origin

Netherlands

Little is known of the Holstein's early years, but some records mention black-and-white cattle existing in the northern Netherlands before the 17th century. In the 18th century, some theories suggest, the British imported Holsteins to improve their dairy cows. The first U.S. imports, by Dutch settlers in New York, were simply called "Dutch" cattle. A newspaper editor renamed the breed Holstein-Friesian, which was later shortened to Holstein.

The U.S. breed society formed in 1871, after the Holstein had already become well established. The original cattle were a dual-purpose breed; this changed on American farms. The farmers focused solely on the trait in which the Holstein already excelled: milking. This emphasis, and some selective breeding, set in motion the process toward today's cud-chewing milk machine.

Black Baldy

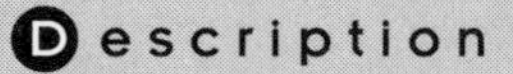

A WHITE-FACED, BLACK COW.

Taking cues from its forebears, the Black Baldy shows a white face and a black body. On some animals, depending on the coloration of the parents, white spots on the body or eye patches may occur. The cross, like the Angus, is polled. Like the Hereford, the Black Baldy has a large, stocky, rectangular body and a fairly straight topline.

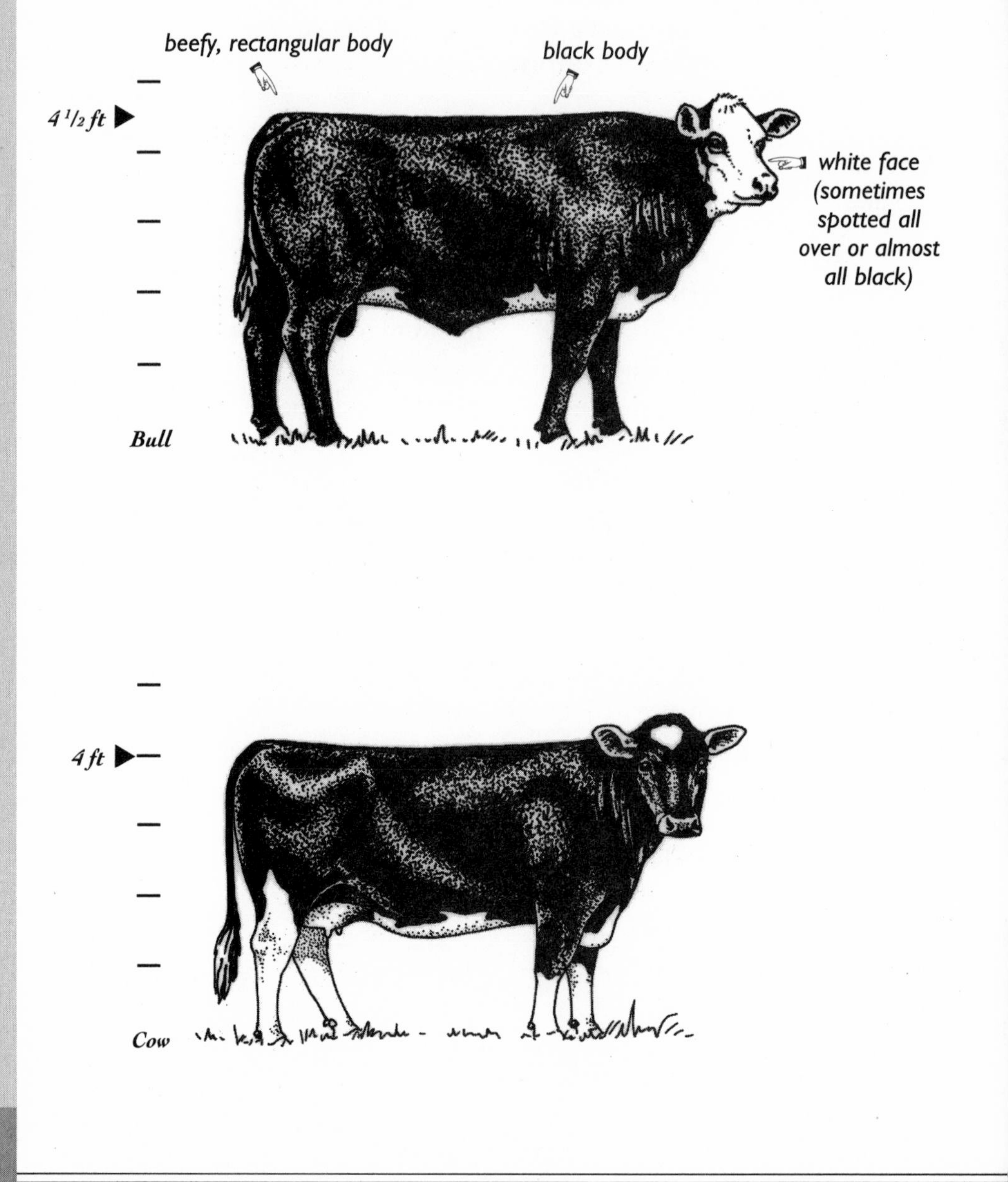

black with white face

Black Baldy

- **Synonyms:** ***Black Whiteface***
- **Distribution:** ***Ubiquitous***
- **Registrations:** ***None***
- **Bull's Average Weight:** ***1,200 pounds***
- **Cow's Average Weight:** ***1,100 pounds***

Purpose

Beef

The cross of the white-faced Hereford and the all-black, polled Angus creates an offspring with superior hybrid vigor: that is, faster growth and greater bulk (the offspring are 10 percent heavier) than the parents; but the benefits only last for one generation. A continual supply of purebred or top-grade parents is needed to breed Black Baldies. The two particular parent breeds are used not only for their availability but also because the Angus has high quality beef and the Hereford grows rapidly. The Black Baldy has decent cold-weather tolerance; even the calves are hardy.

Origin

The Black Baldy is not considered a breed in its own right, since it is raised only for the first generation's hybrid vigor; still, it is common enough throughout the country to almost have breed status. Farmers probably first started crossing the Hereford and the Angus when the two breeds were both available and in enough supply that no one had to worry about diminishing the stock of one breed or the other. For the actual breeding, the Angus is more often used as a sire since the Hereford cow has slightly better milking and mothering ability than the Angus cow. Any farmer anywhere can breed this type of cattle; it is seen throughout the United States.

Per day, a cow spends 6 hours eating and 8 hours chewing cud.

Description

A WHITE-BELTED COW.

A white belt encircles the body behind the shoulders and before the rump; the udders are usually white. According to the Dutch Belted breed society, a truly belted cow has a belt at least 6 inches wide fully surrounding the cow. Overall conformation continues the breed's traditional dual-purpose shape: not too angular but with the udders still big. Physical features include a slightly dished face, black tongue, smooth coat, and small stature. Other cows that are sometimes belted include the Galloway, the Welsh Black, and the Swiss Brown.

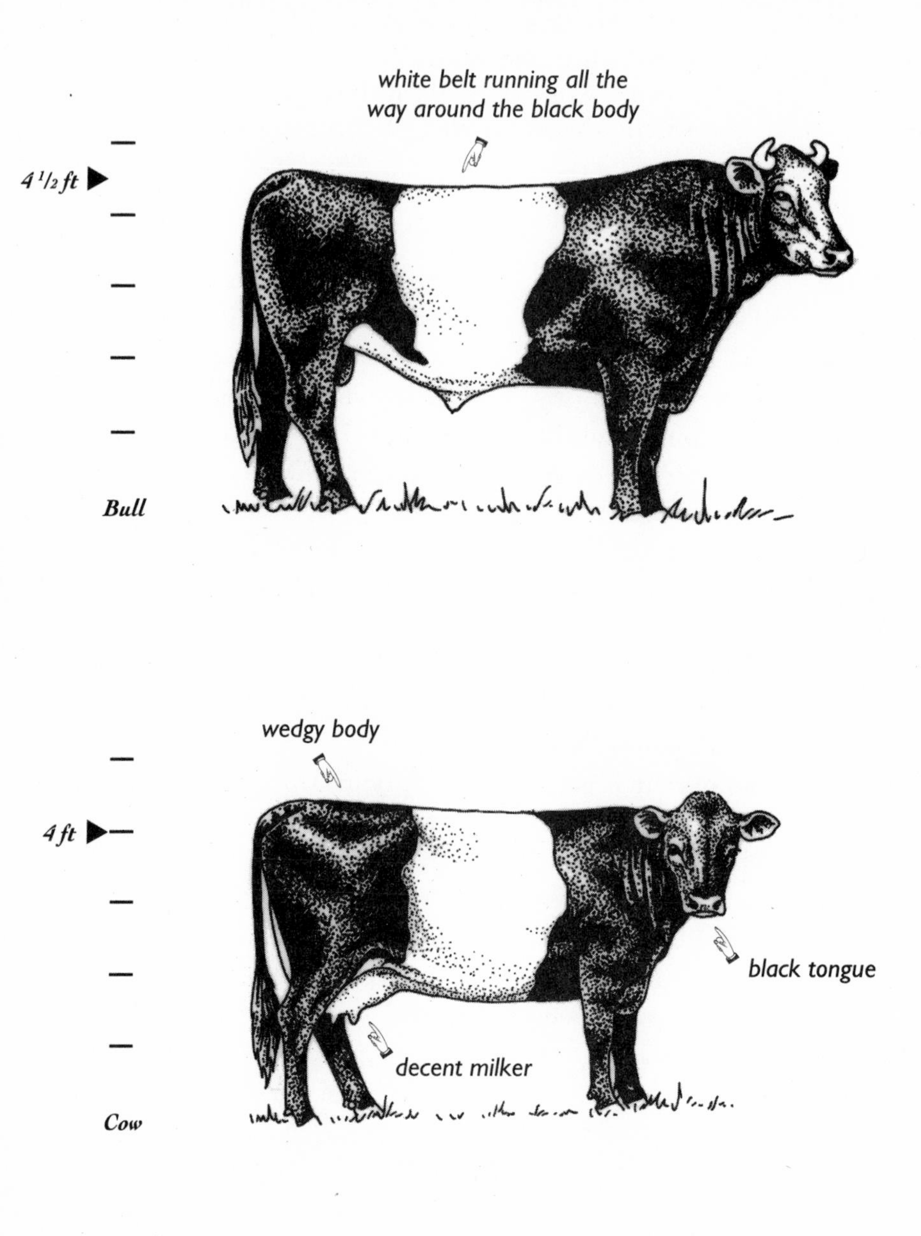

black & white

Dutch Belted

- Synonyms: ***Dutch Belt, Lakenvelder, Gurtenvieh***
- Distribution: ***Critical***
- Registrations: ***about 50***
- Bull's Average Weight: ***1,980 pounds***
- Cow's Average Weight: ***935 pounds***

Purpose

Dairy

Originally a dual-purpose dairy and beef breed, today the Dutch Belted is often raised just for its cool appearance as well as its decent milking characteristics. The cows are said to be easy handling and docile, and are persistent milkers. The Dutch Belted is noted as being ideal for pasture based farms, those farms where the cows eat more grass than grain.

Origin

Netherlands

The Dutch Belted, like the Holstein, has a fuzzy history prior to the 17th century. Only a few records and some paintings showing belted cows grazing on the estates of Dutch nobility remain from these early years. Historical interpretation has it that the cows, though never too numerous, probably grazed on peasant farms as well. During this time, farmers more than likely crossed the cows with any other nearby breed, including the then similarly shaped Holstein. Only later, when the Dutch Belted was imported into the United States starting in 1838, were the cows truly purebred. One of the Dutch Belted's first importers was P. T. Barnum. He presented the cow with an emphasis on its royal heritage and with a big show flair as "The Cow of Kings." His belted cattle helped save the breed from getting crossed out of existence: after retiring from the ring, the cows were put out to pasture (a good thing in this case), and continued to be purebred. Recently, some descendants of P. T. Barnum's cattle were even sent back to the Netherlands to help the original stock. The Dutch Belted society formed in 1886.

Description

THE SHAGGY, LONG-HORNED COW.

The most common color is some shade of black or brownish black, but the range encompasses dun, yellow, silver, white, red, and brindle; no one color is favored. Whatever the color (and there may be more than one in a herd), it's always solid. The Highland's two most noticeable traits are the long, shaggy hair, which covers the body from the ears to the tail, and the only thing not hidden behind all that hair—the long, intimidating horns. Other features include short legs, small ears, and a large foretop that covers up the face.

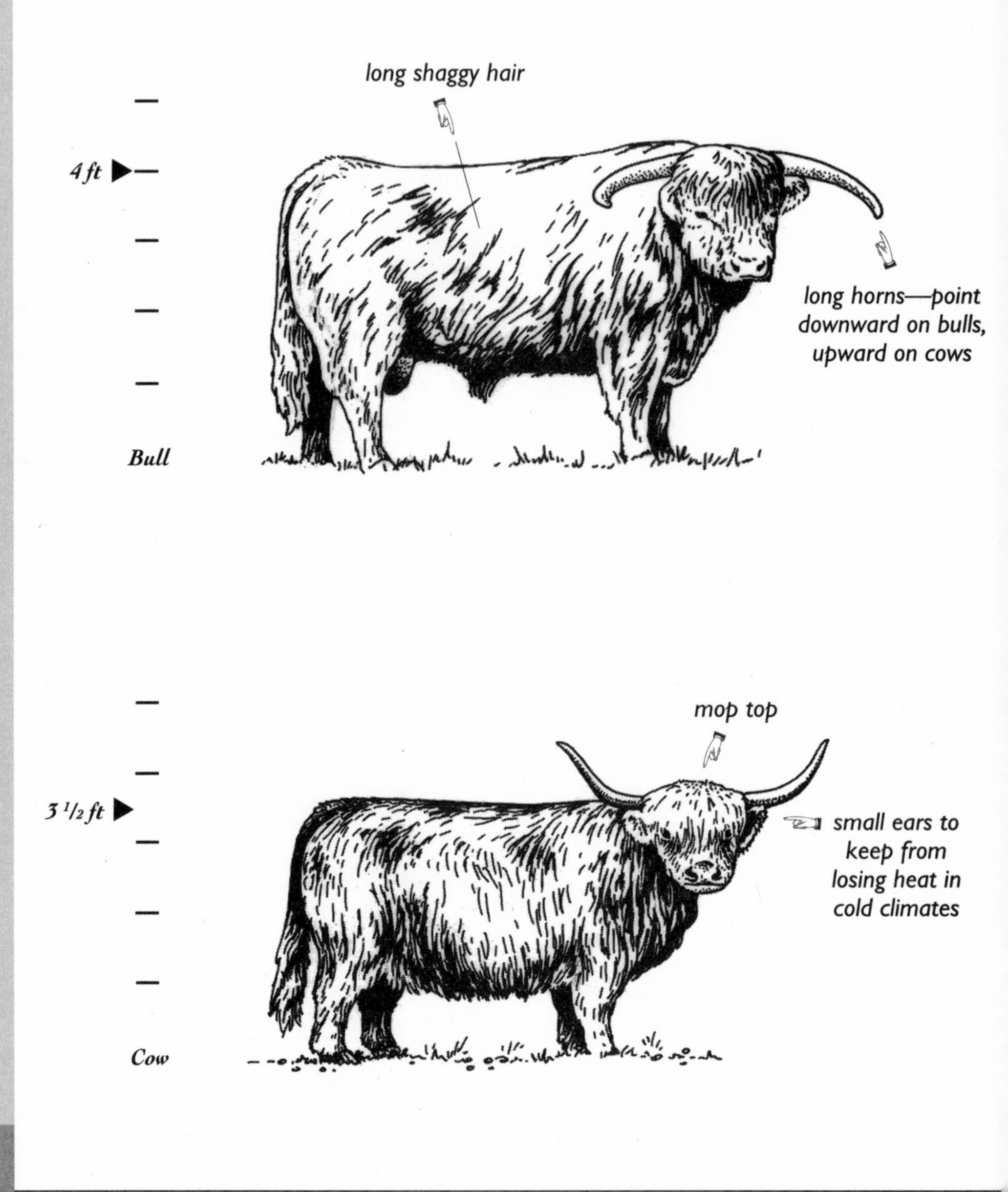

long-horned & black

Highland

- Synonyms: ***Kiloe, Kyloes, West Highland, Scotch-Highland***
- Distribution: ***Sparse***
- Registrations: ***over 1,000***
- Bull's Average Weight: ***2,280 pounds***
- Cow's Average Weight: ***1,100 pounds***

Purpose

Beef, draft, and hairy, horned aesthetics

The feature farmers most appreciate is the Highland's hardy, durable character. The cows can basically be let loose on the range; fearless of the conditions, they will keep on chewing cud and tending calves. This type of worry-free ranching works especially well in the colder climates, where the Highland's long coat keeps it warm. The coat gives the breed an added feature of extra lean beef, since the Highland doesn't have to deposit fat subcutaneously to keep warm, like other cows. The Highland's beef always ranks as one of the best in terms of flavor.

Origin

Northwest Scotland

Sketchy references to Highland cattle go back to the 12th century, but some researchers believe the breed existed even at the time of the British Iron Age. Its hairiness and long horns proved necessary for survival in the rough solitude of northwest Scotland's islands and mountains. The cow's isolation ended with the coming of the industrial revolution; the big cities' increased demand for beef ushered in the time of the Highland's greatest popularity. To meet this demand, cattle drives of up to 30,000 head went south to the English markets. On these journeys, the herds had to cross whatever happened to be in their way, including mountain passes and the ocean channels. These open-water channels, called Kyloes, presented such a daunting challenge to the cows' endurance and character that "Kyloes" became the breed's alternate name. In 1922, the Highland was imported to Montana. The breed has an enthusiastic following worldwide. The U.S. Highland breed society organized in 1948.

Description

A SLEEK, POLLED, BLACK COW.

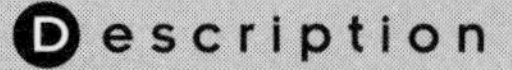

Tradition has the Angus colored solid black, though dabs of white might occur on the udder. The breed also has a genetically recessive red color trait, which means that a red cow may appear amongst an otherwise all-black herd. (A separate Red Angus line also exists.) The Angus has a classic British beef shape, with a barrel-like body, stocky legs, and a smooth, sleek coat. Other features are a fairly short face, small poll, and quick, lively step. The Angus has never had any horns, and still doesn't.

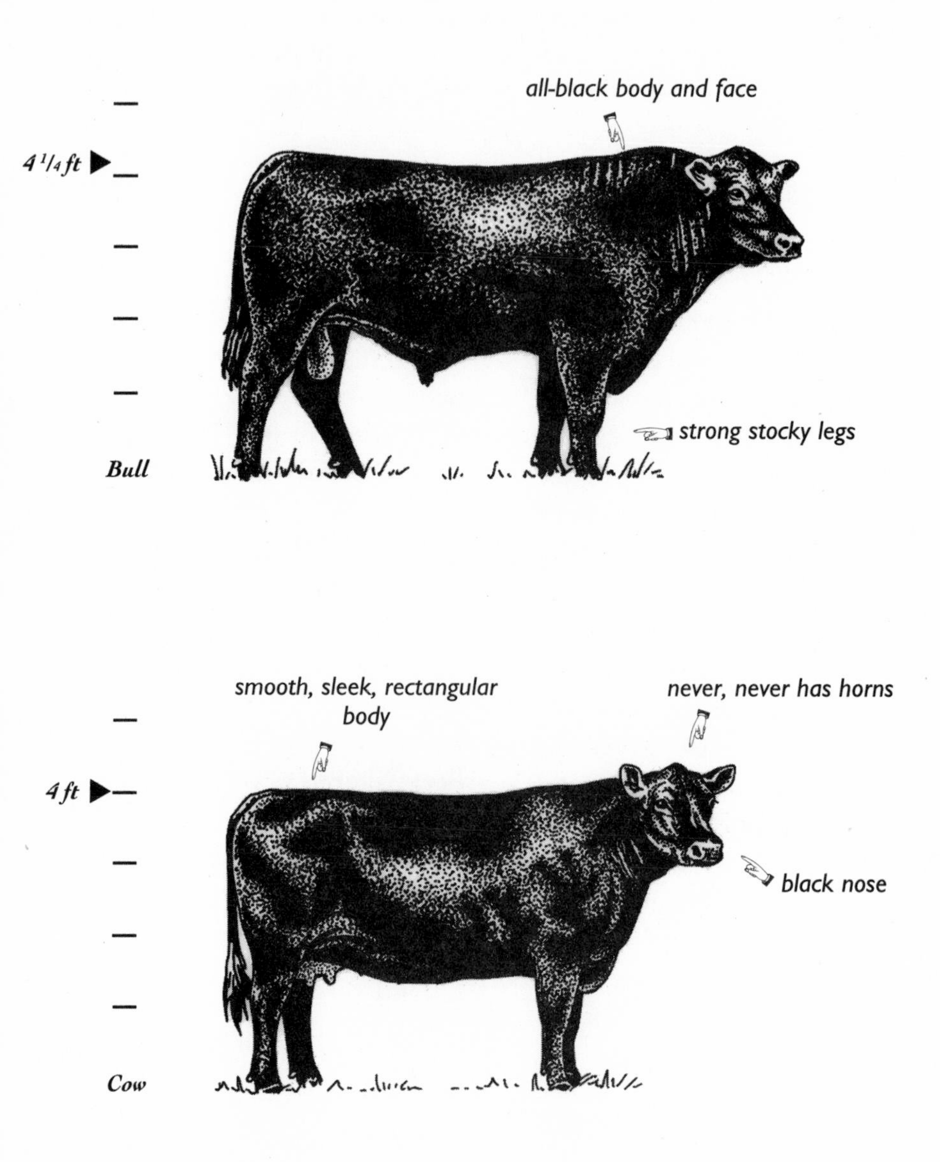

polled & black

Aberdeen-Angus

- Synonyms: ***Angus, Polled Angus, Northern Scotch Polled***
- Distribution: ***Ubiquitous***
- Registrations: ***about 180,000 (includes Red Angus)***
- Bull's Average Weight: ***2,000 pounds***
- Cow's Average Weight: ***1,150 pounds***

Purpose

Beef

The Angus has the finest quality beef of the industrial breeds. USDA studies confirm that its beef consistently ranks first in taste, texture, and color. Taking advantage of this fine quality, the Angus Breed Association promotes its product by marking the prime cuts "Angus Beef," which makes it one of the few beef cuts labelled with its source. Other noted characteristics include quick weight gain, early maturity, and excellent hybrid crossing vigor.

Origin

Aberdeen and Angus Counties, Scotland

The breed started during the late 18th century with the crossing of Angus Doddies and the Buchan Humlies. Found in the Scottish counties of Angus and Aberdeen, these two polled cattle lines most likely originated from the local indigenous cows as well as from other cattle found in the region, such as the Highland. Development, refinement, and setting of the beef traits came at the end of the 18th century. The breed was formally recognized in 1835 and imported to the United States in 1873 (a breed society formed ten years later).

The first imported cattle quickly found favor with farmers and ranchers around the country. Western ranchers especially appreciated the Angus's traits and frequently crossed it with the then-plentiful Texas Longhorn. This cross resulted in a polled cow that retained the longhorn's ruggedness but was easier to ship eastward in the cramped cattle cars and that had an improved beef quality, up to the eastern market standards. Today, the Angus ranks as the second most popular beef cow.

Description

A SHAGGY, BLACK, POLLED COW.

The common color is brownish black, but the breed comes in a multitude of shades, including red, dun, white belted, and white with sharply contrasting black points. The cow's shagginess actually forms a two-part weather protection system: part one is the long outer hair, which repels wind and rain; part two is the short, thick, furry underhair, which keeps the cow nice and warm. Fortunately, the Galloway sheds its coat in the summer. Other features include a broad muzzle, a broad forehead, a stocky build, and an overall shape similar to the Highland's.

black

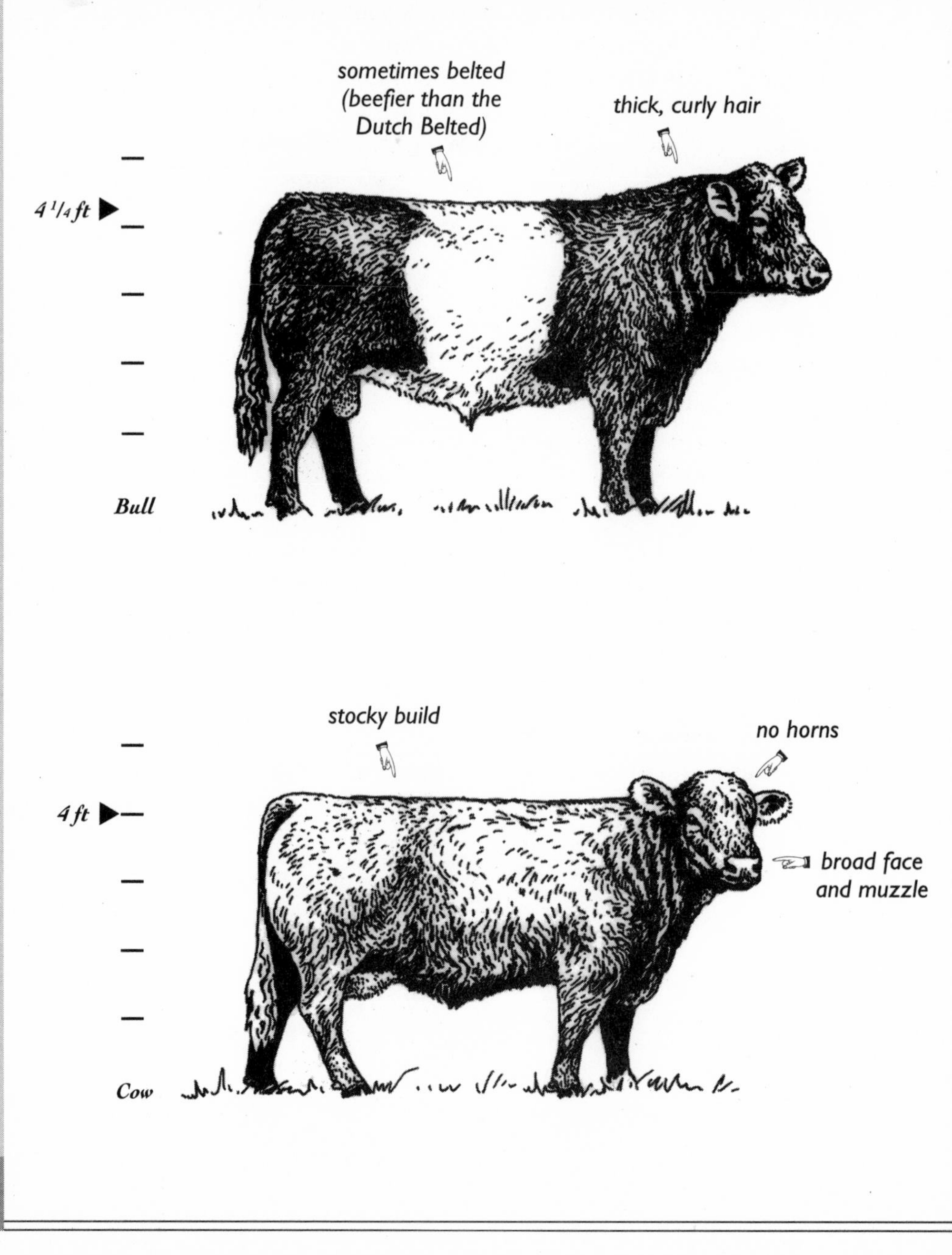

Galloway

- Synonyms: ***Southern Scots Polled***
- Distribution: ***Rare***
- Registrations: ***about 350***
- Bull's Average Weight: ***1,790 pounds***
- Cow's Average Weight: ***1,265 pounds***

Purpose

Beef, dairy, and furry aesthetics

Because of the Galloway's decent milking ability, it started off as a dual-purpose breed; today it focuses mostly on beef production. The breed is especially noted for its hardiness and superior mothering ability. One tale highlighting these qualities tells of cows, in defense of their calves, chasing off hungry wolves. Other traits of the Galloway are its easy manageability, its disease-free genetics, and its ancillary products, which include great hides.

Origin

Galloway, Scotland

The Galloway originated in the province of the same name in southwest Scotland. Although the breed's exact chronology is uncertain, one plausible theory connects the Galloway with its northern neighbor, the Highland. The idea for this relationship stems from the two breeds' close proximity over the ages and from their similar body conformation. One vague reference was made in the 11th century to polled cows, which would mean the two breeds diverged at some point before this period. More accurate sales records that mention the breed exist only from the 18th century. The Galloway herdbook formed in 1862 along with the Angus herdbook, and a separate breed society formed in1878. The Galloway was imported to the United States by way of Canada in 1853, and a U.S. breed society started in 1882.

Cow Fact:

It takes about 1 1/2 gallons of milk to make 1 gallon of ice cream.

Description

A MEDIUM-HORNED, JET BLACK COW.

The usual color is black, but it also comes in dark chocolate brown, white, belted, smoky, lineback, and blue. The unique blue color arises from the cross of a black animal with a white animal, which gives an offspring with hairs that are all mixed up and sort of bluish in color. On the lighter-shaded cows, the black pigment shows up as black points. Additionally, the Welsh Black has bright eyes; short legs; soft, long hairs; and long, black-tipped horns. The breed has an amiable appearance, kind of like a big, friendly dog.

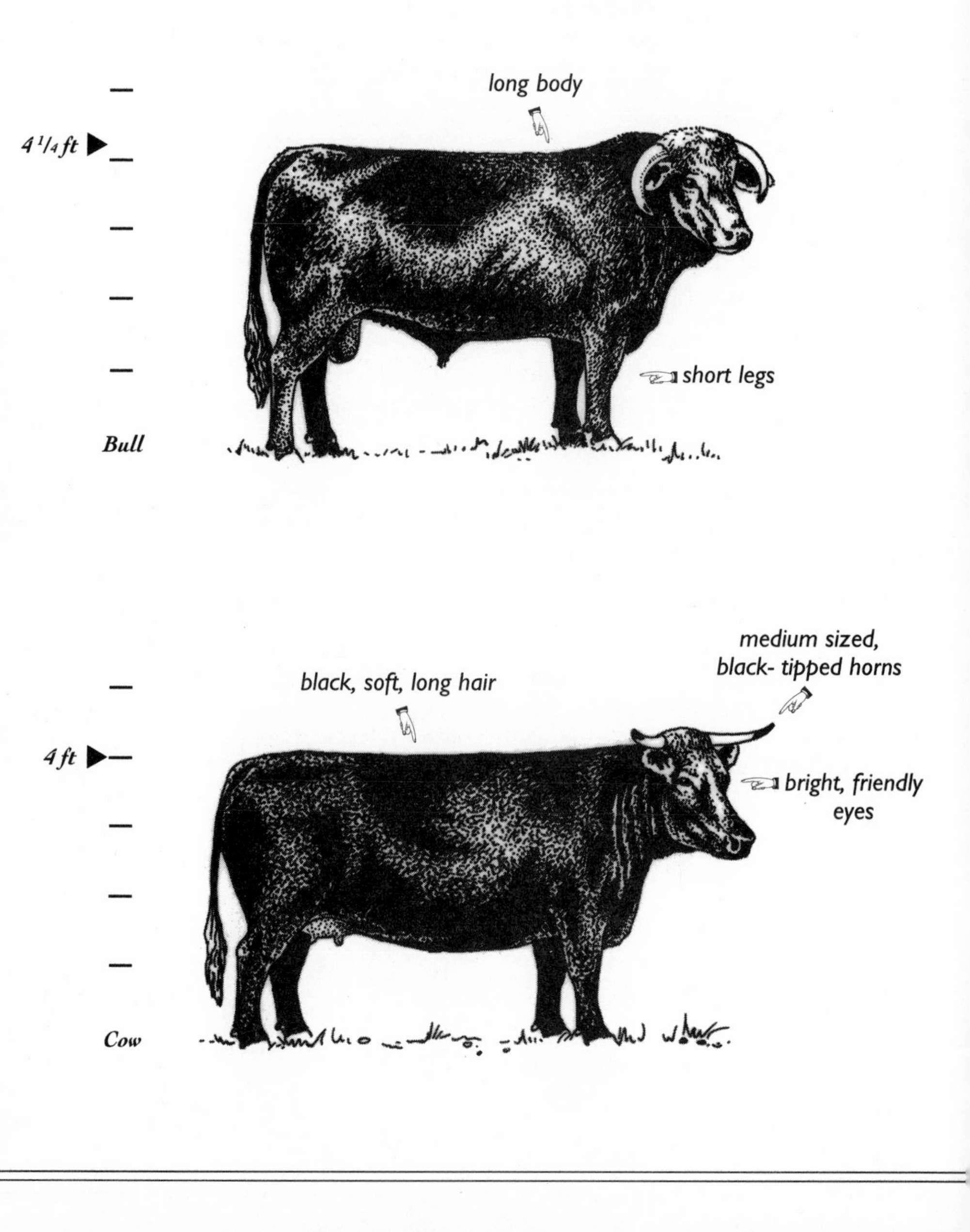

black with medium-length horns

Welsh Black

- **Synonyms:** ***Gwartheg Duon Cymneig (Welsh)***
- **Distribution:** ***Limited***
- **Registrations:** ***about 1,000***
- **Bull's Average Weight:** ***1,900 pounds***
- **Cow's Average Weight:** ***1,200 pounds***

Purpose

Beef and dairy

The Welsh Black is noted as a good homestead cow because of its gentle disposition, its healthy calves, and its decent milk yields. Its milking tradition has earned the cow the title, at least with some ranchers, as the brood cow breed.

Origin

Wales, United Kingdom

The Welsh Black originated in the late 19th century from a consolidation of two ancient breeds: the Anglesy and the Pembroken. The Anglesy was a northern mountain breed known for its long, beefy body, and the Pembroken was a southern dairy breed that was said to gain weight quickly when dry. Anglesy and Pembroken are both considered "Celtic" cattle, an ancient type of Brachyceros or shorthorn cattle first raised in Mesopotamia. A herdbook was started in 1873 but was later broken up into two regions; in 1904, the two regrouped. The society chose black as the preferred color, though it allowed any color. The Welsh Black was imported to the U.S. in 1963. The U.S. Association formed in 1975.

Cow Fact:

Total 1994 value of the 100,110,000 head in the U.S.A. was $63,016,005,000.

Description

THE SMALLEST COW.

A solid shade of brownish black or, at times, dark red. The most apparent feature is its size; it is the smallest of all cattle breeds. A mature Dexter stands half as tall as a full-grown Holstein, and weighs half as much. Still, the breed is fairly well proportioned overall, so that it might be mistaken for a larger cow when seen alone—that is, with no one standing near it to reveal its true petiteness. Other features include black skin pigmentation, a short face, a broad forehead, a hairy foretop, and short horns.

black and small

"This cow is short!"

Dexter

- Synonyms: ***Dexter-Kerry***
- Distribution: ***Rare***
- Registrations: ***about 600***
- Bull's Average Weight: ***1,000 pounds***
- Cow's Average Weight: ***800 pounds***

Purpose

Dairy, beef, and draft

The Dexter is a modern-day triple-purpose cow. Daily dairy production works out at about 2 gallons of milk and 3 quarts of cream—a lot of milk even after the calf gets its share; more than enough for a large family's daily use. The sides of beef also come in family-sized proportions: just the right size for the average household freezer. And finally, though small in stature, the Dexter is said to be a strong plow puller and a tractable animal.

Origin

Southwest Ireland

The two theories about the Dexter's origin both have the Kerry, a rare Irish breed, as a major source. One idea is that the farmers of the rural Irish county of Kerry selected progressively smaller Kerry cattle until they formed a miniature version of the cow. The other theory claims that the Dexter started from small mountain cattle that were crossed with the Kerry. The Dexter started to become known in the 18th century. The peasant population saw to its first husbandry, since their small plots required its small size and hardy nature. Later, the novelty of the small, cutesy cows made the breed popular on English country estates. The United States breed society formed in 1911. At around the same time, the Dexter was frequently being crossed with the Kerry. This crossing became so common that the cows were often called Dexter-Kerry. Today, the breed is popular worldwide.

Cow Fact:

Today dairy cows produce double the amount of milk compared to cows from the 1960s.

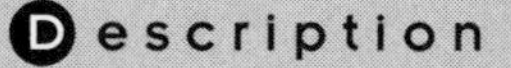

Description

A DARK, BROWNISH BLACK MILKING COW.

The shiny, dark-colored coat is highlighted by a lighter shade underneath, a yellow-striped muzzle, and, occasionally, a striped back. The Canadienne's excellent dairy conformation looks similar to that of the Jersey, but the Canadienne has a larger body, a straight face, and a consistently dark body coloration. Other physical characteristics include dark skin pigment, black hoofs, short horns, short hair, and a refined, stately appearance.

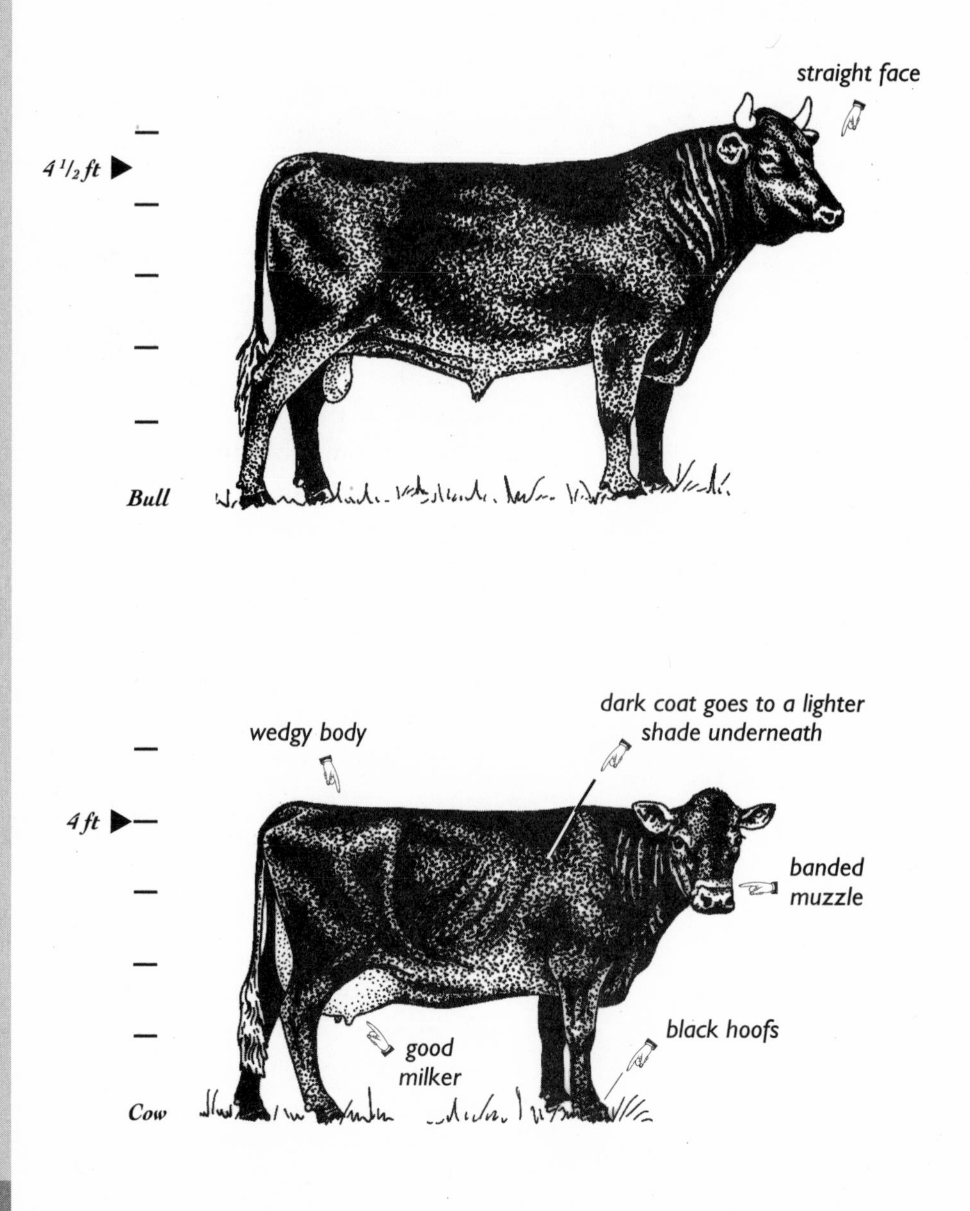

black

Canadienne

- **Synonyms:** ***Canadian, French, Quebec, or Black Jersey***
- **Distribution:** ***Critical***
- **Registrations:** ***about 70 in Canada***
- **Bull's Average Weight:** ***1,790 pounds***
- **Cow's Average Weight:** ***1,210 pounds***

Purpose

Dairy

The Canadienne is a perfect cow for extensive dairy production—that is, on any farm where pastures, rather than expensive grains, provide the major part of the cow's diet. The cow's qualities include a gentle nature, high fertility, a long lifespan, and good milk production for its size. The Canadienne also adapts well to hilly terrain and tough, invigorating climates.

Origin

Quebec, Canada

In 1541, Québecois Jacques Cartier imported some French dairy cattle; they were some of the first cows on North American soil. These first Canadian cows were probably related to the Normande, Jersey, or Alderney cattle that were then present around coastal France. About sixty years later, more cows arrived in Quebec by way of Samuel de Champion, providing a strong enough foundation to keep the breed going for the next couple of centuries. During those centuries, farmers on the relatively isolated Québecois farms used no formal breeding techniques, allowing natural selection to shape the breed into the region's number-one dairy cow. All the years of adaptation, and the breed itself, nearly ended when a crossbreeding craze struck. In 1886, the Canadienne breed society formed; its enthusiasm for the breed saved the day. In the 1960s, the Canadienne again faced trouble, this time from the government's milk promotion program. Farmers were forced to emphasize milk quantity over the Canadienne's traditional forté: quality. The easiest way for them to do this was to crossbreed, leading to further reduction in the number of purebred Canadiennes. Fortunately, in the 1980s, the government saw the light and provided assistance for the breed's continued survival.

Corriente

Description

A STRONG, SMALL, HORNED COW.

Any color and all color combinations are permissible expect a solid white with no skin pigmentation. The Corriente has a small, narrow, lean stature; though not beefy, it has a great deal of strength and endurance. The shoulders and rump are high set. The girth is deep, though the dewlap is small. Other characteristics include medium-length, curved horns that are strong and heavy, and a broad, triangular-shaped face.

multicolored with medium horns

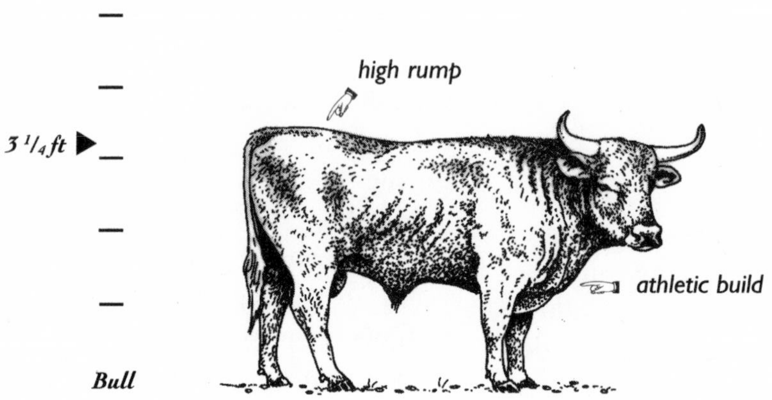

high rump

3 1/4 ft

athletic build

Bull

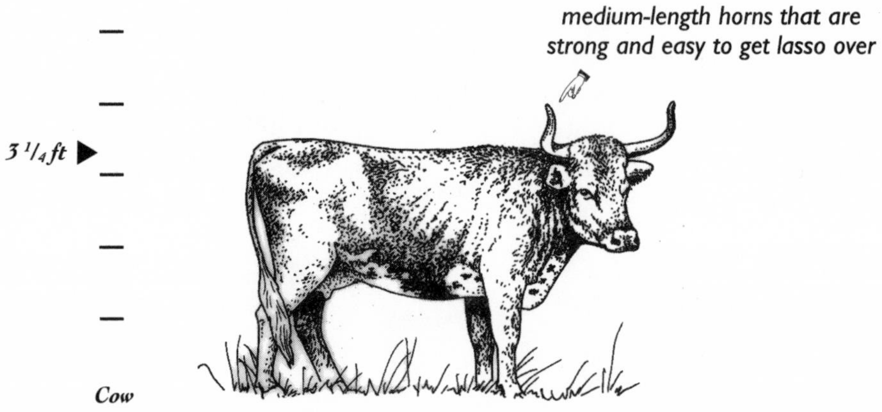

medium-length horns that are strong and easy to get lasso over

3 1/4 ft

Cow

Corriente

- Synonyms: ***Criollo and Chinampo (in Baja, Mexico)***
- Distribution: ***Limited***
- Registrations: ***N/A***
- Bull's Average Weight: ***1,000 pounds***
- Cow's Average Weight: ***800 pounds***

Purpose

Rodeo

Because the Corriente combines all the characteristics necessary for the rodeo (small size, stamina, agility, a gentle nature, and well-attached horns—traits that most modern cattle breeds have lost), it is considered the classic cow for roping and bulldogging in the ring. It also adapts well to the hot, dry climate of the Southwest, and has excellent foraging ability; disease resistance; a long, productive life; calving ease; and range awareness (intelligence). The cattle can supposedly go 2 to 3 days without water and are considered browsers, like deer, not grazers.

Origin

Mexico

Spanish settlers in Mexico and the Caribbean originally imported Spanish cattle, the precursors to the Corriente, during the 15th century. Though the breed itself never gained widespread popularity, the early cows probably had a big influence on the Texas Longhorn. Only a few Corriente survived to more modern times; these were found in the remote regions of Mexico and in select parts of the southern United States. The surviving cows were raised mostly on small subsistence farms and in close proximity to people; this is apparent in their gentle disposition. The breed is actually said to be friendly, in contrast to other semiferal breeds like the Longhorn or the Highland. Today, the Corriente could be considered one of the few land race cattle breeds of North America, the others being the Canadienne and maybe the Texas Longhorn. The name comes from northern Mexico, where *corriente* refers to any native cattle found in the countryside. The North American Corriente Association formed in 1982.

Description

A LONG-HORNED, RANGY COW.

A spotted, dull red is the most common color, but the cattle are said to come in all patterns and colors. The long horns, usually lyre-shaped, make the perfect Cadillac hood ornament; an average spread is around 5 1/2 feet while record lengths span up to 9 feet. The breed is built overall for ranging and rustling, which is apparent in the long legs, high-set shoulders, and easy stride. Since the cows no longer get the exercise of cattle drives, their conformation is changing to a more elliptical and beefier look.

multicolored with long horns

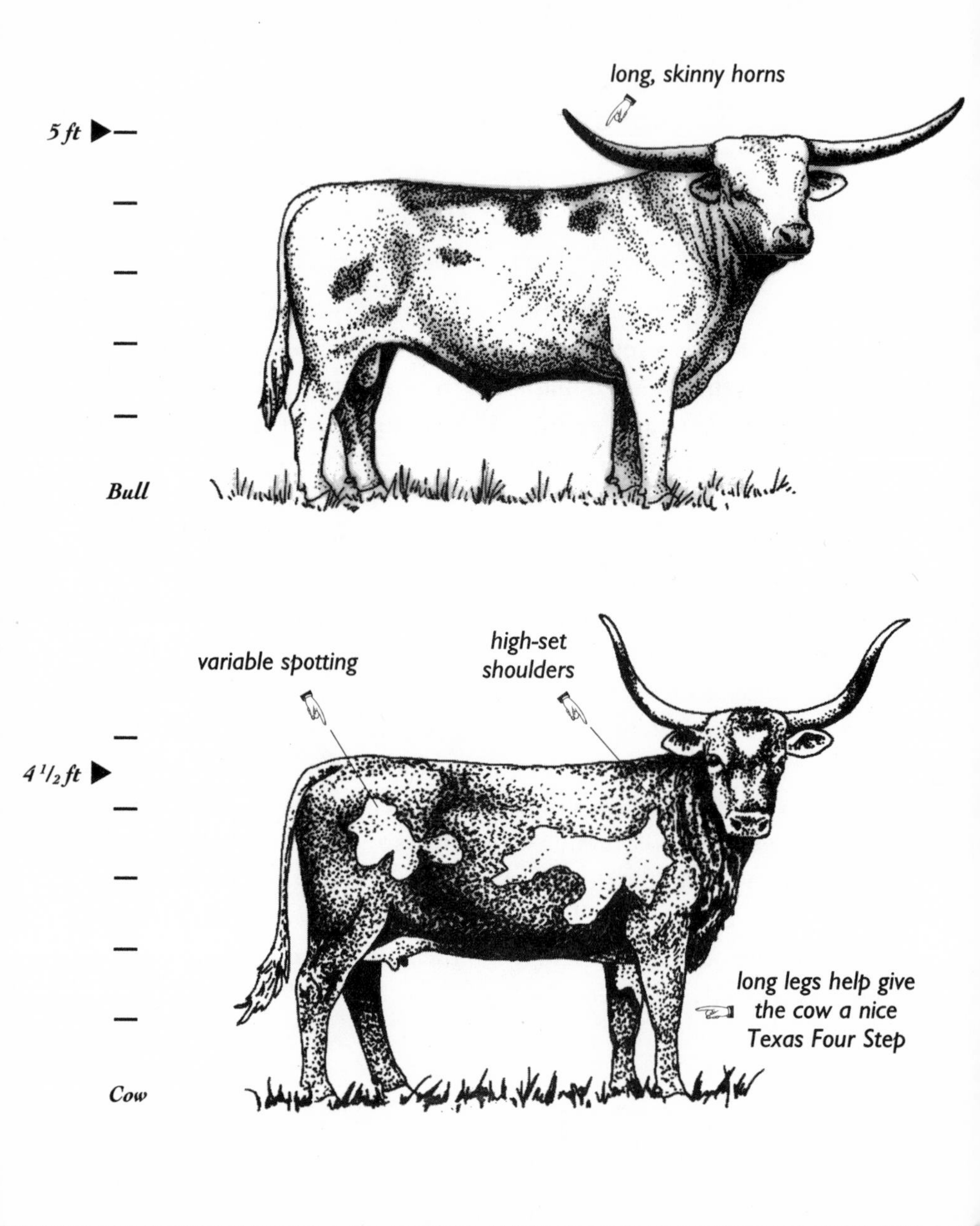

Texas Longhorn

- Synonyms: ***Longhorn***
- Distribution: ***Common***
- Registrations: ***over 12,000***
- Bull's Average Weight: ***1,430 pounds***
- Cow's Average Weight: ***770 pounds***

Purpose

Beef and ropin' 'em dogies

Ranchers these days probably raise the Texas Longhorn as much for the breed's hardiness as for its cool traditional look. This innate hardiness and the cow's adaptability allow cowboys to use a more extensive system: the cattle can get along just fine without a lot of pampering on those big Texas ranches. Some of the Longhorn's rough-and-ready qualities include rustling ability, hot-weather tolerance, lean beef, good mothering ability, and disease resistance.

Origin

Texas, U.S.A.

In 1640, vaqueros drove the first Spanish cattle north to Texas. Some of these cattle escaped, turned feral, and slowly took the place of the Bison on the wide open range. The cow's conformation probably changed in the late 19th century, when farmers started settling the area. Some theories have it that the Texas Longhorn took its long horns and varied coat patterns from the English Longhorn. Because of a demand for beef in the huge east-coast markets, the Texas Longhorn's numbers increased and reached a high point of 5 million around the turn of the century. Their decline began with a couple of hard winters and was compounded by the crossing in of other newly popular British breeds. In 1929, the total population of Texas Longhorns reached such a low point that the government stepped in to save the breed. Yet, lack of interest left the Texas Longhorn's numbers at only 1,500 until the mid-1960s, when its popularity started to grow slowly. The breed association formed in 1964. Today, ranchers from Texas up to Minnesota have rediscovered the breed, and it numbers about 100,000 head nationwide.

Description

THE BIGGEST HORNS.

The most common color is red-and-white skewbald, but all colors and patterns may occur, including black, brown, yellow, dun, gray, brindle, and white. The horns are the biggest of any breed (sorry, Texas), their spread often reaching up to 8 feet and their bases a fat 8 inches in diameter. Other traits include long legs, a slanting rump, a small thoracic (shoulder-area) hump, and a long, insect-switching, ropelike tail.

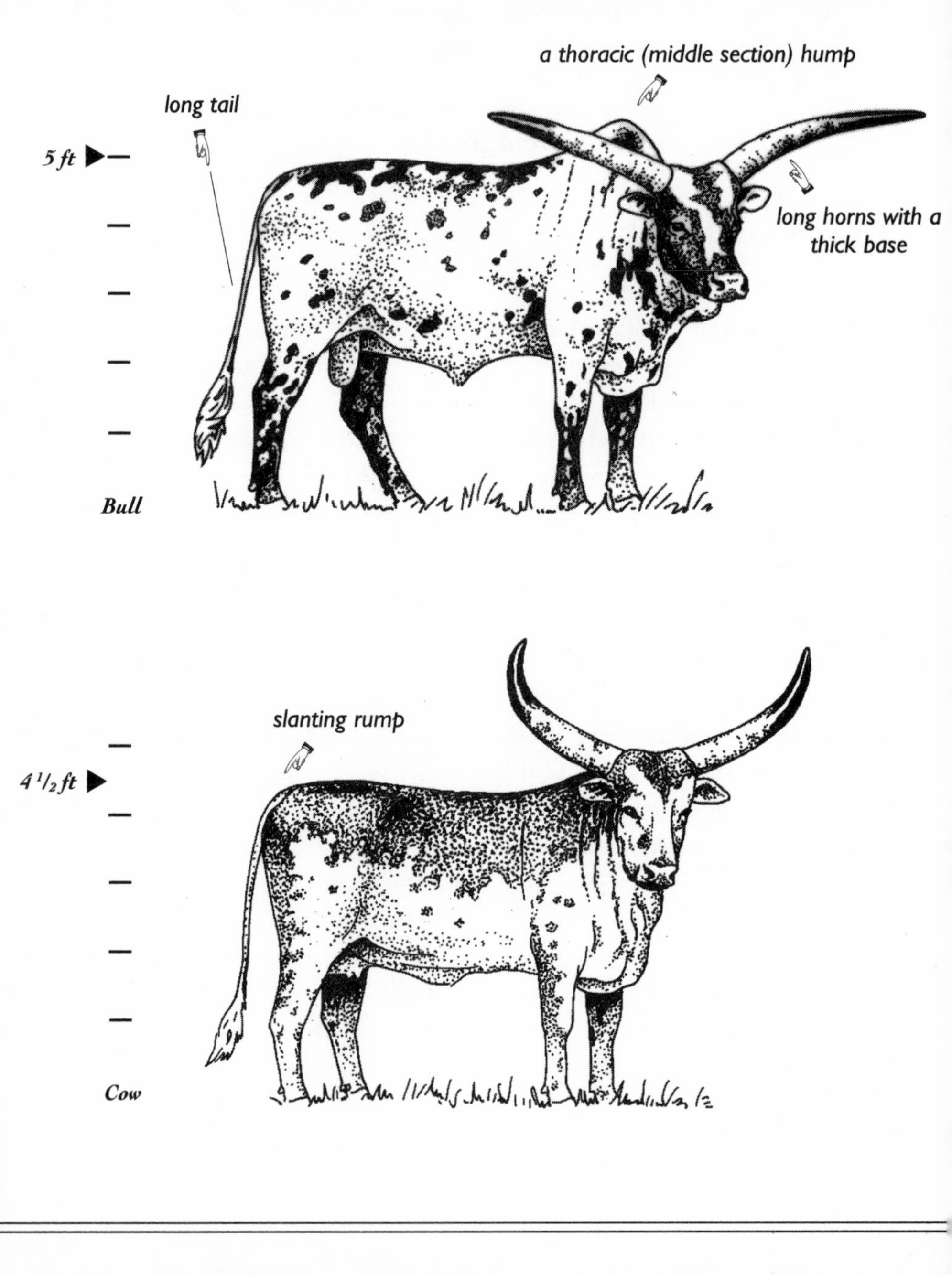

multicolored with long horns

Watusi

- Synonyms: ***Ankole-Watusi, Burundi, Kivu, Rwanda, Inkuku, and King Cattle***
- Distribution: ***Critical***
- Registrations: ***about 30***
- Bull's Average Weight: ***2,000 pounds***
- Cow's Average Weight: ***1,500 pounds***

Purpose

Long-horned aesthetics

Most ranchers who raise the Watusi do it for one reason only: the huge horns. Partly, ranchers may want to augment the hood ornament market; but these magnificent cattle are appreciated more when wearing their own horns. The Watusi has some other excellent characteristics that could be useful on ranches, such as rangeability, insect tolerance, hot-weather adaptability, and drought resistance. Part of the animal's ability to withstand drought comes from its digestive process, which extracts almost all the water found in plant matter; the cows have dry dung. Another quality that makes for easy ranching is that Watusi prefer to herd together closely; even at night, they form a tight-knit sleeping group called a glum.

Origin

Rwanda, Burundi, and Zaire

The Watusi comes from the Ankole, a sanga type of cattle native to the central African region that includes Burundi, Rwanda, and Zaire. Over the years, ranchers from this area developed the long horns and the interesting color patterns on the coats using the same selective breeding techniques as their European cattle-raising counterparts. Still, the cow's main purposes were milk and beef production; the physical features had only aesthetic importance. The Watusi was imported to the United States in the 1960s by way of Sweden. Today, though the cows are not too numerous, they have found some popularity as an exotic breed and a rodeo attraction.

Cow Fact:

New Zealand is first in butter consumption at 26 lb/person/year. The United States spreads on 4.6 lb/person/year.

Description

A WHITE-FACED COW.

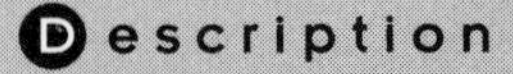

A dark red body contrasts with the white face, underside, chest, and switch. A completely white face is traditional, though some breeders prefer that red encircle the eyes. The Hereford has a blocky body; stocky legs; straight topline; long, level rump; broad face; and short, thick horns (not found on the polled line). The Hereford differs from the other white-faced cow, the Simmental, in that it is smaller, has—in general—less spotting, and is darker red in body color.

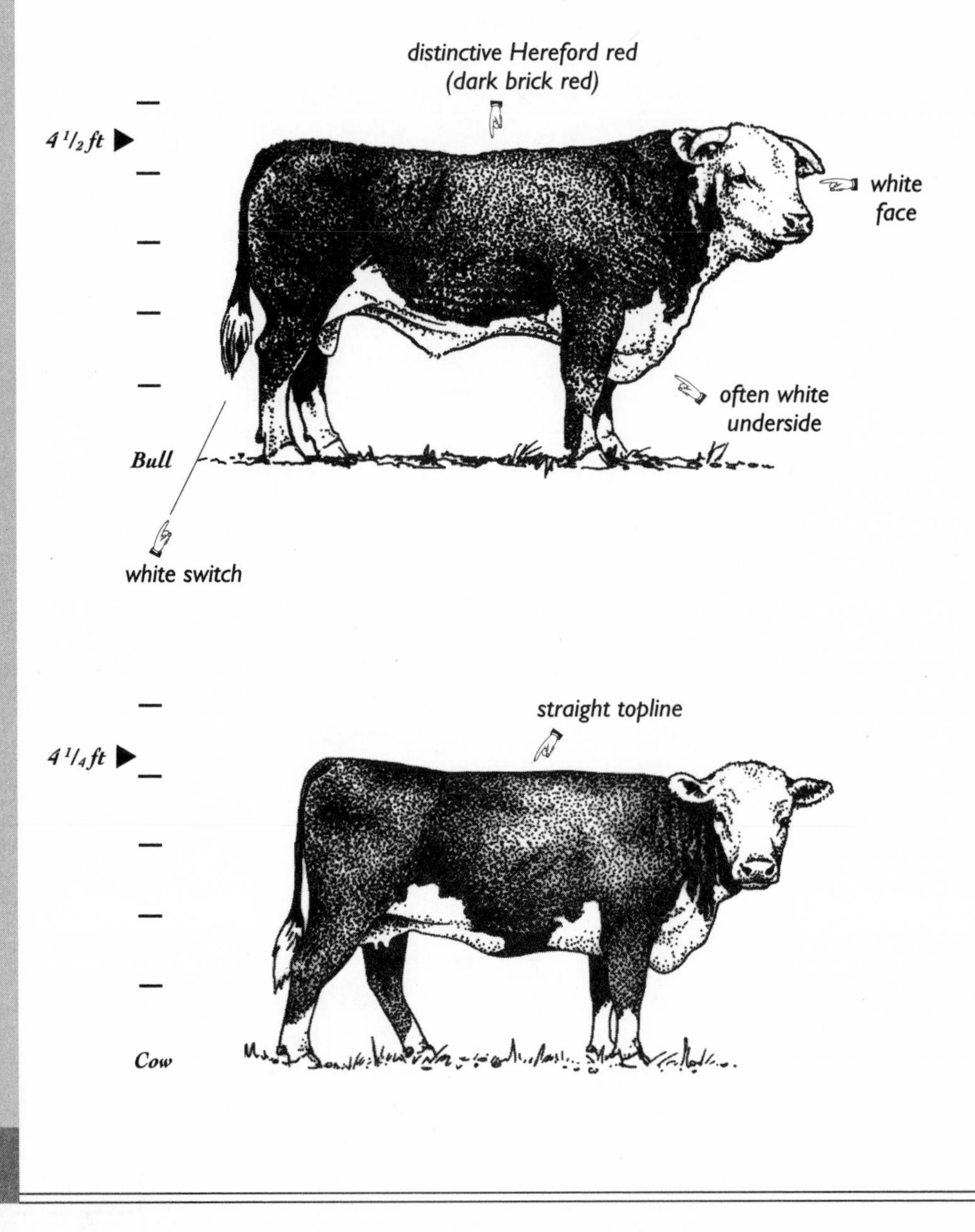

white face and red body

Hereford

- Synonyms: ***White Faced and Herefordshire***
- Distribution: ***Ubiquitous; including polled Herefords***
- Registrations: ***over 200,000***
- Bull's Average Weight: ***2,200 pounds***
- Cow's Average Weight: ***1,540 pounds***

Purpose

Beef

Historically, the Hereford was a triple-purpose breed, but with changing times the main focus has turned to beef production. The Hereford's qualities include fast growth, docility, good beef quality, and decent cold-weather tolerance. Because of a lack of pigmentation around the eyes and short eyelashes, the cows may have some problems with the sun, especially in the bright southern states.

Origin

Herefordshire, England

The Hereford started off in this shire just west of London, in 1742, as a fat beast of over 3,000 pounds (one humongous bull tipped the scales at 3,640 pounds). The breeder Benjamin Tomkins first realized the Hereford's market possibilities, and emphasized early maturity, shorter legs, and good beef quality; he disregarded the cow's color. His breeding stock were local cattle, so, like the neighboring Sussex and Devon breeds, the early multiple-purpose Hereford was a completely red cow. Not until the early 19th century did breeder John Hewer set in place the identifying white face. The first Herefords to make it to the United States arrived in 1847. The British herdbook formed in 1846 and the United States herdbook started in 1881. In the late 19th century breeders created a separate polled line. Today, this polled line is as popular on farms as the horned line, though it's hard to tell the two apart since ranchers often clip the horns of the naturally horned line. The Hereford can be seen nearly everywhere: as well as being the most popular beef cow on American farms, it's the most widely distributed breed in the world.

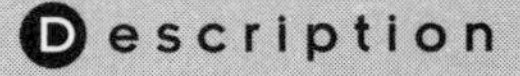

Description

A WHITE-FACED, YELLOWISH RED COW.

The traditional look has a white face and some white underneath (Hereford-like), but the breed places no restrictions on color or pattern. Other fairly common color variations and patterns include solid white, solid red, solid black, and spotted skewbald. Additional physical characteristics are a long face; a larger head than is found in the British breeds; a long topline; a beefy rump; a loose, hanging dewlap; and large ears set low on the head.

white face and red body

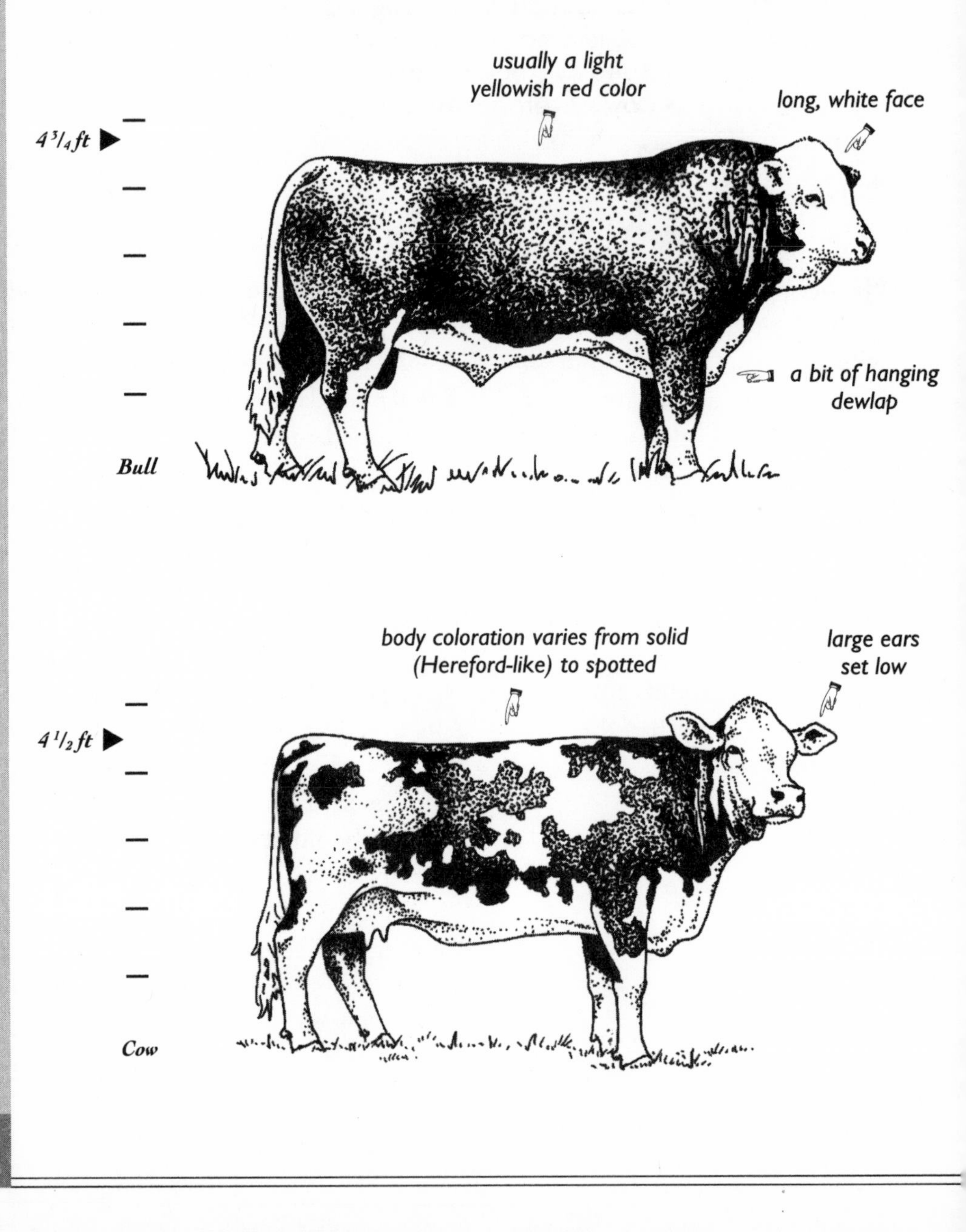

Simmental

- Synonyms: ***Fleckvieh (German); Pezzata Rosa (Italian); and Pie Rouge, Montbeliard, and Abondance (French).***
- Distribution: ***Ubiquitous***
- Registrations: ***over 90,000***
- Bull's Average Weight: ***2,350 pounds***
- Cow's Average Weight: ***1,650 pounds***

Purpose

Beef and dairy

In former times, the Simmental was a true triple-purpose breed, and in some remote European villages it is still used for animal traction. Today, the breed is broken up into different beef and dairy lines. American farmers primarily use the beef line, but they can still take advantage of the breed's latent dairy traits in crossbreeding. Some of the Simmental's qualities include good mothering ability, a docile disposition, decent quality beef, strong bones, and ruggedness.

Origin

Switzerland

Simmental is the German name for the Simme Valley, one of the valleys where the cows were raised and developed. In the predevelopment-preindustrial days of the breed, when there were still many family farms, each Alpine valley could claim its own individual line. New mountain roads put an end to the cow's isolation and to its distinct varieties: crossbreeding became possible across the passes. This crossbreeding unified and standardized the cow's characteristics. In 1862 the breed was formally recognized. Of the various lines used in these across-mountain-pass crosses, the most popular proved to be the white-faced Bernse. In the 1960s and 70s, some Red and White Holstein blood entered the line. Today, the Swiss cow population is 50 percent Simmental. The Simmental Breed Society and herdbook started in 1890. The cow was imported to the United States in the 1880s, but the United States Breed Society did not form until 1968. The breed has an estimated 40 to 60 million members worldwide. In the United States, it ranks as the third most popular beef breed.

Pinzgauer

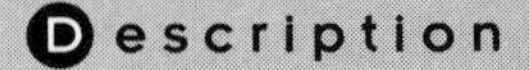

Description

A BROWN-SIDED COW.

The chestnut brown markings on this unusually colored cow cover the head and chest, narrow down on the sides, and end before the rump. Orange pigment adds color around the eyes and udders. Other features include a long frame; short, black-tipped horns; a large chest; and muscled hindquarters. The Pinzgauer may be confused with the rare Lineback or the even rarer Randall Lineback, which are also breeds with white backs and brown sides.

brown sided with white

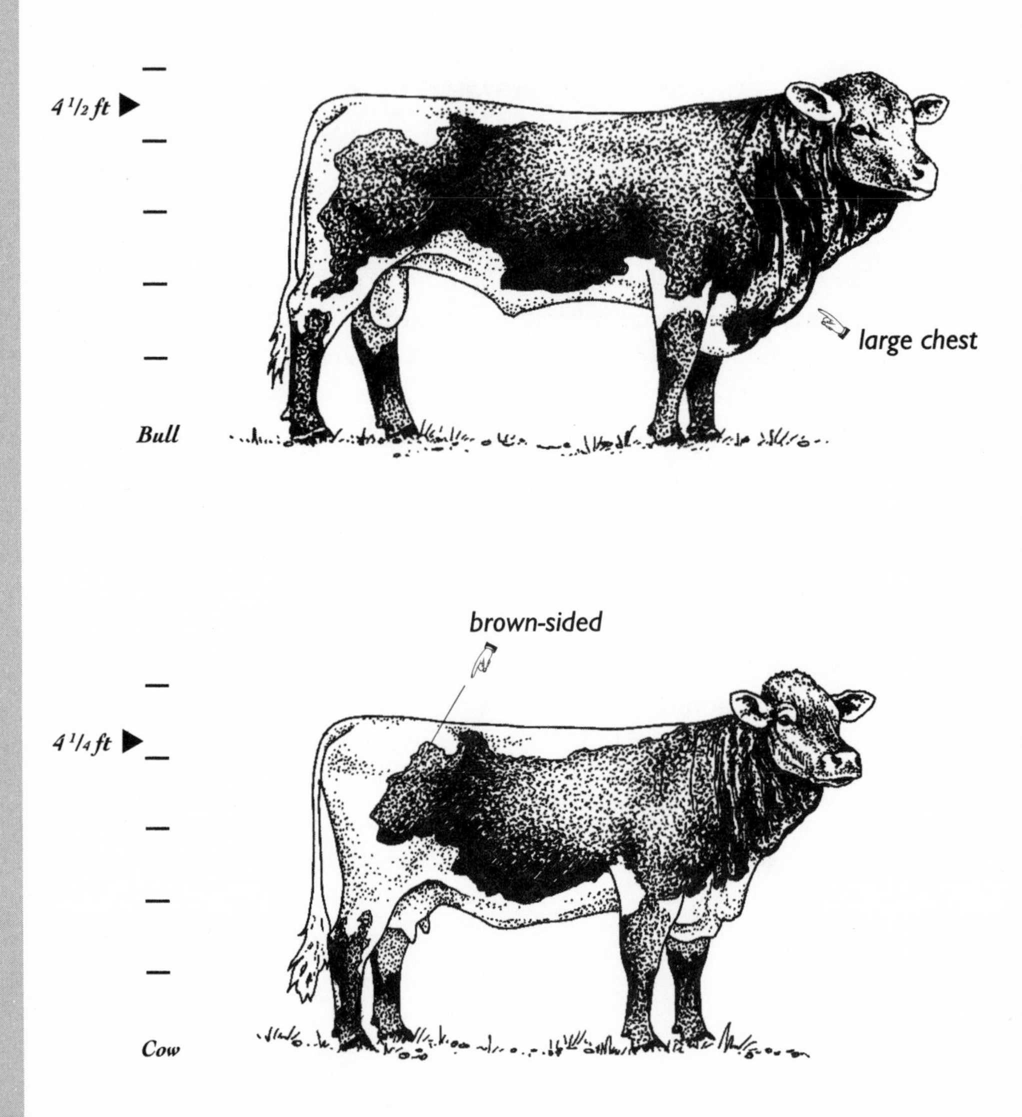

Pinzgauer

- **Synonyms:** *Pinzgau*
- **Distribution:** *Limited*
- **Registrations:** *about 1,000*
- **Bull's Average Weight:** *2,550 pounds*
- **Cow's Average Weight:** *1,475 pounds*

Purpose

Beef and dairy

The cow's former draft occupation made it a triple-purpose breed, but today it is a dual-purpose cow in Europe. Europeans also prize the breed for its high-quality leather, which they use for shoe soles. In the United States, it is most often a beef cow. The Pinzgauer is said to impart excellent heterosis, to adapt to tough climates, and to have a friendly nature. Plus, the cows have rich milk, which fattens the calves up quickly.

Origin

The Alps

The Pinzgauer originated in the Austrian Pinz Valley; from there, its range extended into the bordering Alpine regions of Italy and Germany. The breed arose from a crossbreeding of Valais cattle with indigenous Austrian cattle in the middle of the 19th century. Recently, some Red and White Holstein blood was also mixed into the breed. The unusual coloration still shows; this tells us that the cow has escaped major cross-breeding and other breeding manipulations, since the lineback pattern is more prevalent on land race breeds. Today the Pinzgauer's numbers are decreasing in Austria (it makes up only 13 percent of all cattle) because of the pressure of the industrial stocks. A U.S. breed society formed in the early 1970s, though its present situation is unknown.

Cow Fact:
Hay ranks third, behind pasture and corn, as the most popular livestock feed.
1
2
3

Description

A MASKED, TRICOLORED COW.

The two colors making up the spots are brown and yellow-red; the background is white (the third color). The spots can take on any pattern; they also usually encircle the eyes to give the cow an interesting masked appearance. The muzzle, the ears, the hoofs, and the skin under the spots are also darker-colored. Other physical features include a dished face (or, as the French say, *coup de poing*); fine, short horns; a large, clean chest; decent body depth; a wide pelvis; a long frame with a straight topline; and big udders.

red, brown, & white

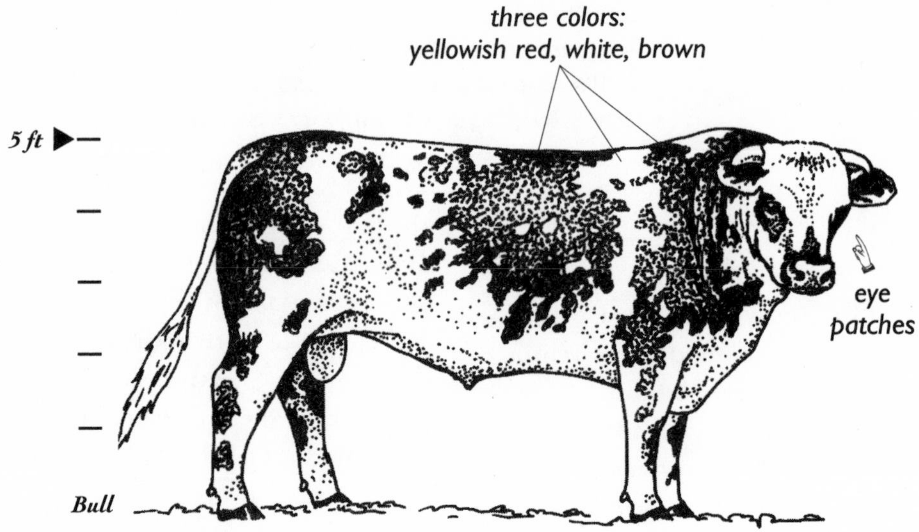

Normande

- Synonyms: ***Normandy and Norman***
- Distribution: ***Limited***
- Registrations: ***about 350***
- Bull's Average Weight: ***2,600 pounds***
- Cow's Average Weight: ***1,650 pounds***

Purpose

Beef and dairy

The Normande continues to be an excellent dual-purpose breed to this day. The dual type occurs more frequently in France, where a Normande cow won a world butterfat record earlier this century. American farmers focus primarily on the Normande's beef uses, and let the Holstein handle the milking job. Besides decent milk yields and beef quality, traits include an even disposition, good maternal instincts, a tough constitution, and good cold-weather tolerance (the breed association is located in cold northern Minnesota).

Origin

Normandy, France

The Normande is an amalgamation of the cows that were once common along the coasts of France. Viking cattle (left behind by the Vikings in the 9th and 10th centuries) and the ancient, cool-named Isgny were some of the early cows incorporated into the tri-colored cow. Later, between 1845 and 1860, some Shorthorn and Jersey blood found its way into the breed. During World War II, when the Allies made Normandy the focal point of the invasion forces, the region's farms and livestock were almost totally destroyed. The Normande has made a strong comeback in France, and today numbers around 3 million head. The breed has also gathered a following worldwide, with associations in Brazil, Colombia (over 1.6 million head), Madagascar, and Spain. A French breed society formed in 1883. The Normande was imported to the United States in 1974.

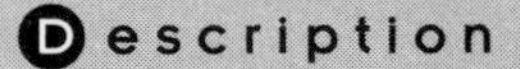

Description

A BIG, BEEFY, RED-AND-WHITE COW.

The Meuse Rhine Yssel looks sort of like a pumped-up Red and White Holstein. Traits in common with its northern neighbor (besides color) include a straight face; a broad muzzle; small, incurving horns; a large frame; and large spots. The MRY, though, has a beefy, fuller shape and is a bit shorter in the legs.

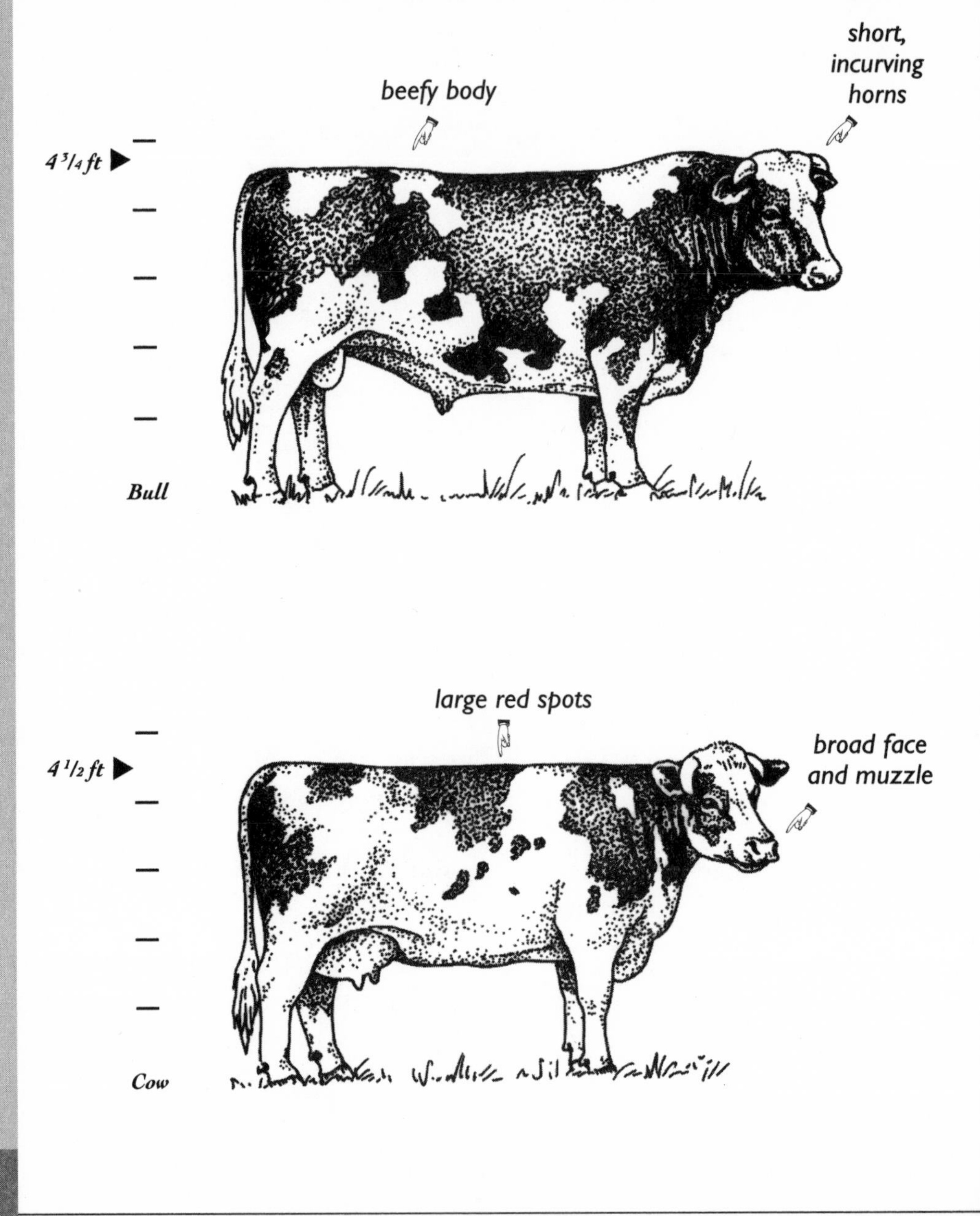

red & white

Meuse Rhine Yssel

- **Synonyms:** ***MRY, Meuse Rhine Issel, MRI, and Roodbont***
- **Distribution:** ***Limited***
- **Registrations:** ***N/A***
- **Bull's Average Weight:** ***2,580 pounds***
- **Cow's Average Weight:** ***1,500 pounds***

Purpose

Beef and dairy

The MRY is a modern-day dual-purpose breed, often used to increase beef value in dairy herds and to boost milk production on beef ranches. The MRY grows fast on concentrated feed, has decent beef quality, and is capable of multiple sucklings.

Origin

Southern Netherlands

The MRY originated in the green triangle—the region between the rivers Meuse, Rhine, and Yssel. During the early years, the cattle in this low-lying region were nearly identical to the closely related Holstein. The two breeds probably diverged when some farmers favored the red-and-white cows over the black-and-white ones. Still, both breeds performed the same double job of supplying milk and beef for the majority of their early years. Only in the last 100 years has the Holstein gone all-out dairy, while the MRY has retained the tradition and the beefiness of the dual. The MRY herdbook formed in 1906, and the cow is one of the most popular breeds in Holland today.

Cow Fact: *From 1866 to 1895 cowboys drove about 10,000,000 cattle out of Texas.*

Cattle could walk between 10 and 20 miles a day. The trails varied in length from 600 to 1,700 miles.

Description

A RED-AND-WHITE COW WITH A WHITE TRIANGLE ON ITS FOREHEAD.

The red-and-white skewbald coloration is similar to the related Shorthorn, though the Maine-Anjou often shows a unique triangular white patch in the middle of the forehead. Black and roan cows are also occasionally seen. Other characteristics include a long body, heavy muscling, and short horns. The Maine-Anjou is the heaviest of the French breeds.

red & white

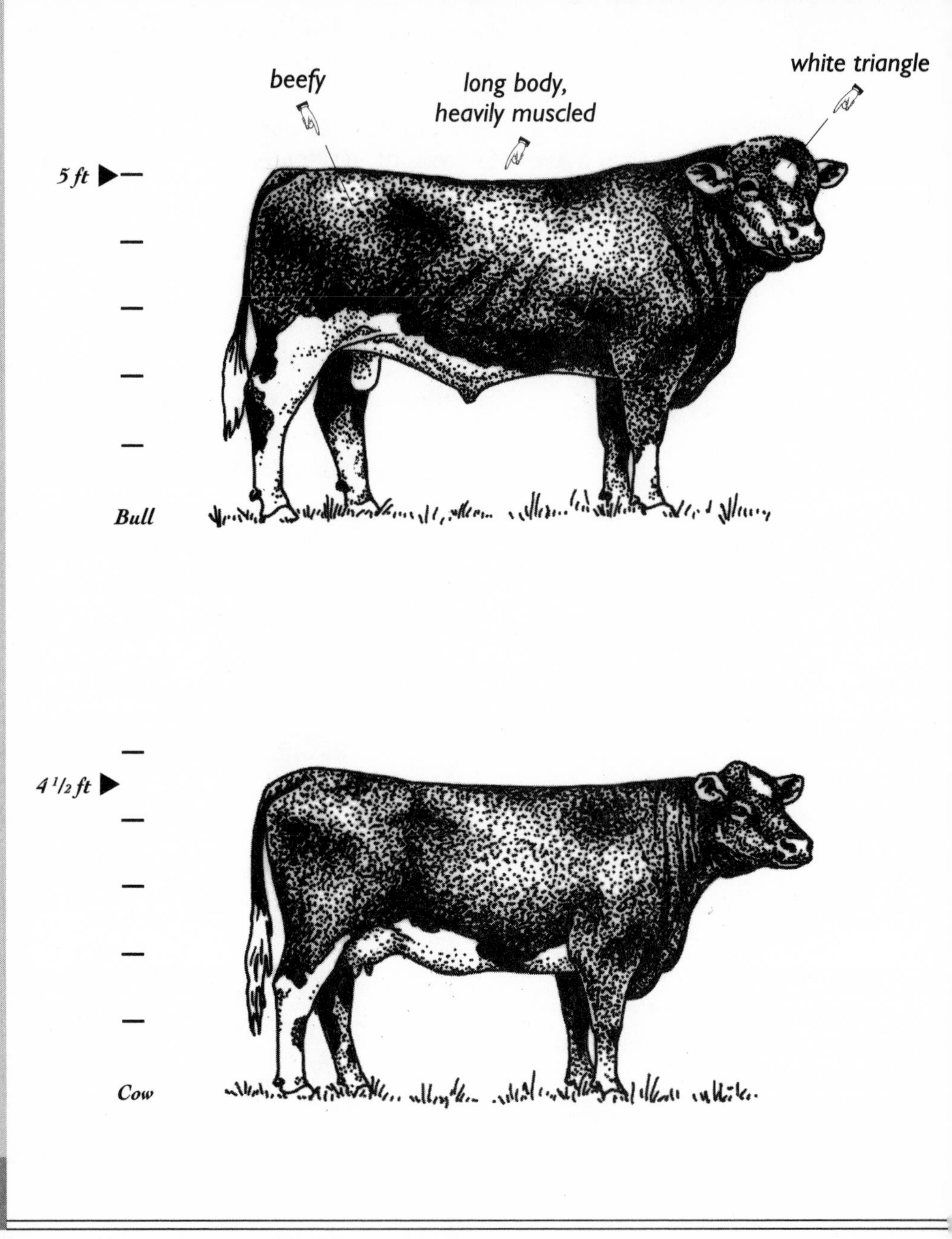

Maine-Anjou

- Synonyms: ***Durham-Mancelle (obsolete name)***
- Distribution: ***Common***
- Registrations: ***about 8,000***
- Bull's Average Weight: ***2,500 pounds***
- Cow's Average Weight: ***1,700 pounds***

Purpose

Beef and dairy

The breed's emphasis is on beef production; it is frequently used to impart its huge size to the offspring in terminal crosses. The cow is noted for a gentle disposition, an adequate milk yield, feeding efficiency, a good growth rate, and an excellent frame.

Origin

Brittany, France

The Maine-Anjou's prior name, Durham-Mancelle, explains the breed's origin. The Durham, an alternate name for the Shorthorn, was crossed with the Mancelle, an old Brittany breed, to form the new and improved breed. The first crossing took place in the 1830s; the idea was to create a dual-purpose breed suited to the region's climate. The Mancelle provided the hardiness and mothering ability, while the Shorthorn, recently improved itself, provided the growth and carcass quality. The cow's current name comes from the northwest French provinces of Maine and Anjou, where farmers have raised the breed throughout the years. Between 1962 and 1970, when the breed was combined with the Armorican, the name was changed to Rouge de l'Ouest (Western Red), but now it's back to Maine-Anjou. The breed was imported to the United States, by way of Canada, in 1969, and the breed society organized the same year.

Cow Fact:
Cows can hear lower and higher frequencies better than humans.
hey, keep it down.

Description

A RED-AND-WHITE, RED, WHITE, OR ROAN-COLORED BEEF COW.

The coloration is variable, with any combination of red and white possible. The roan color occurs when individual red hairs and individual white hairs get all mixed up and make the coat look like a solid color. The cattle have a shorter summer coat and a heavier one for winter wear. Other Shorthorn traits consist of a clean-cut build; a short, broad face; wide-set eyes; and a long, tall frame. Shorthorns do indeed have short, wax-colored horns, which curve slightly with age—at least on those not polled.

red & white

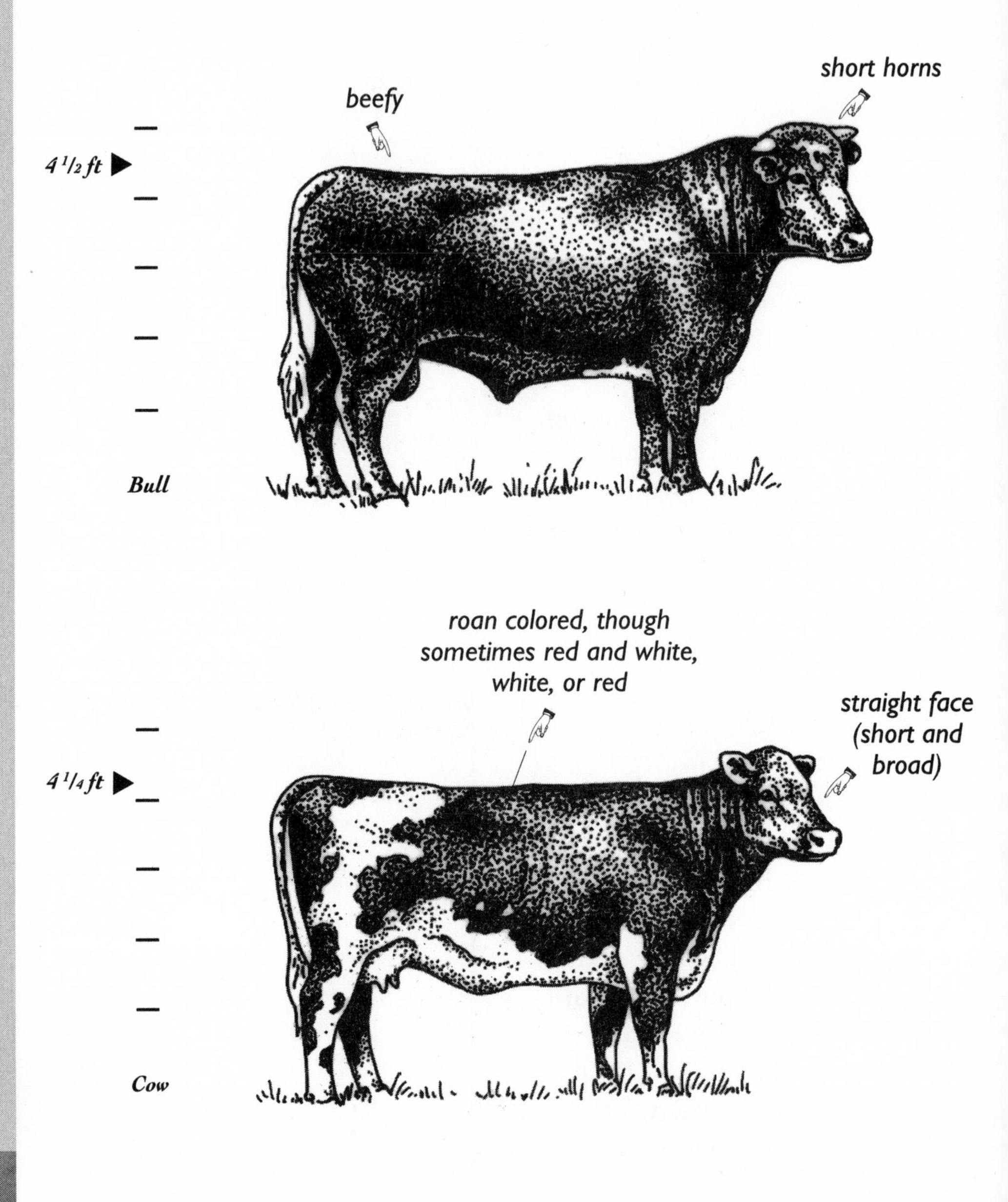

Shorthorn

- Synonyms: ***Durham, Beef Shorthorn, Scotch Shorthorn, and Improved Teeswater***
- Distribution: ***Common***
- Registrations: ***over 11,000***
- Bull's Average Weight: ***1,760 pounds***
- Cow's Average Weight: ***1,100 pounds***

Purpose

Beef

There are three lines of Shorthorn: the beef line, which is the most numerous, the Milking Shorthorn, and the rare Dairy Shorthorn; only the beef and milking are raised in North America. Some of the characteristics of the beef Shorthorn include early maturity, good carcass quality, a hardy constitution, and easy manageability.

Origin

Tees Valley, England

Shorthorn originated from two old cattle types found in northeast England: the Holderness and the Teeswater. The Shorthorn's popularity began in the 18th century; at that time, Robert Collings made methodical improvements using the recently discovered selective-breeding technique (like begets like). The Shorthorn became the first breed purposely developed for improved beef characteristics to have a widespread influence on everyday farms. Another Shorthorn milestone occurred when George Coates introduced the Shorthorn herdbook in 1822, the first of its kind for any cattle breed. The herdbook set minimum cow standards, thereby removing much uncertainty from the job of selecting a good cow. Further development came in the mid-19th century; Scot Amos Cruickshank separated the breed into two lines: the original dual-purpose (now the Dairy Shorthorn) and his new, improved beef Shorthorn. By this time, the original Shorthorn had already made sizeable inroads into the American market; soon, though, the new beef Shorthorn overtook it in popularity. The United States herdbook formed in 1846 and encompassed all the lines, even the polled Shorthorn, which was one of the first U.S.–developed cow lines.

Description

A VARIABLE RED-AND-WHITE MILKING COW.

Cows come in red and white, all red, all white, or roan. The Milking Shorthorn has the same coloration as the beef Shorthorn, though it differs (as the name indicates) in its stronger dairy conformation. The dairy traits include large udders; a wedgy, angular body; and a wide pelvis. The cow has a straight face and buff-colored skin, and is usually polled for the farmer's and the cow's safety.

red & white

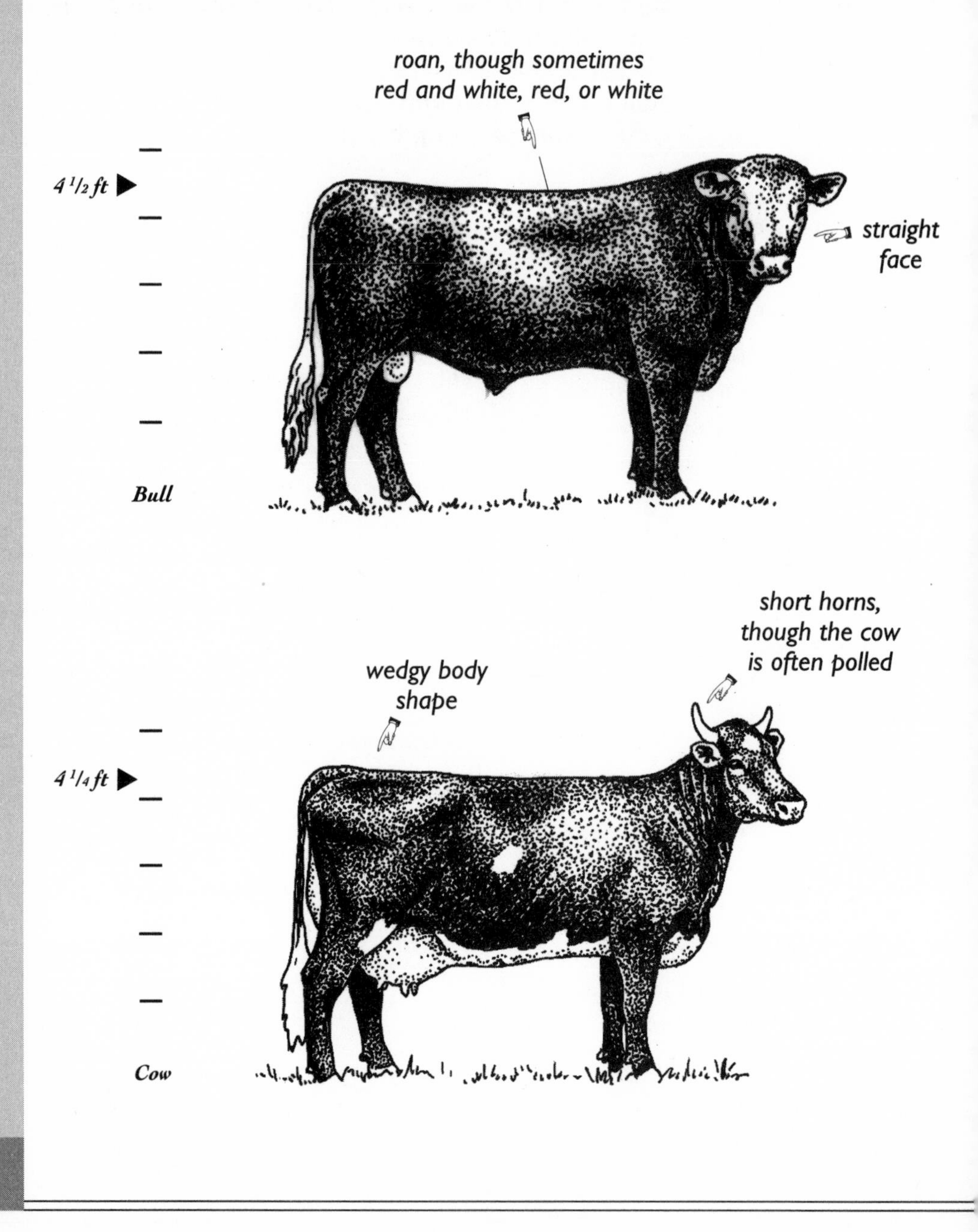

Milking Shorthorn

- Synonyms: ***Dairy Shorthorn (only in New Zealand)***
- Distribution: ***Common***
- Registrations: ***over 3,500***
- Bull's Average Weight: ***2,100 pounds***
- Cow's Average Weight: ***1,400 pounds***

Purpose

Dairy

The Milking Shorthorn is a dairy line, though it is sometimes used in beef herds to increase milk production. The milk yields and milk quality are respectable, but the breed's hardiness makes it shine. The Milking Shorthorn adapts to both hot and cold climates and is considered a good grazer. With 730 milkings in one year, the cow Washita Ann's Bonnie Exp. gave the top Milking Shorthorn milk yield.

Origin

Tees Valley, England

The Milking Shorthorn has the same bloodlines as the beef Shorthorn: it originated in northeast England from the dual-purpose Shorthorn (see Shorthorn). The cow's popularity spread throughout much of Europe and America, but the Shorthorn was superseded by the new Scottish Shorthorn in a crossbreeding craze. During this tumultuous time, those farmers still raising the original dual-purpose cow started emphasizing its dairy qualities, thus making it primarily a milk cow. In 1885, Thomas Bates, looking for greater milk production, further refined the dairy characteristics. Later breeders crossed in some Illawara blood (the Illawara was an old Shorthorn cross from Australia) to further increase milk production. All the Shorthorn lines belonged to the same herdbook until 1944, at which time a separate book was formed for the American Milking Shorthorn.

Cow Fact:

A cow can live 25 years.

aaah sonny-
back in my day
disco was king.

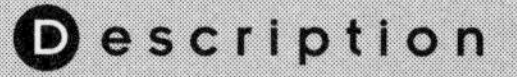

Description

A BIG, AYRSHIRE-LIKE COW.

The color is a deep red, with or without white spots. When there is spotting, it occurs mostly underneath and on the udders in the form of either large or small splotches. Overall, the Norwegian Red has a good dairy conformation: a wedgy body, a well-attached udder, and a nice, refined look.

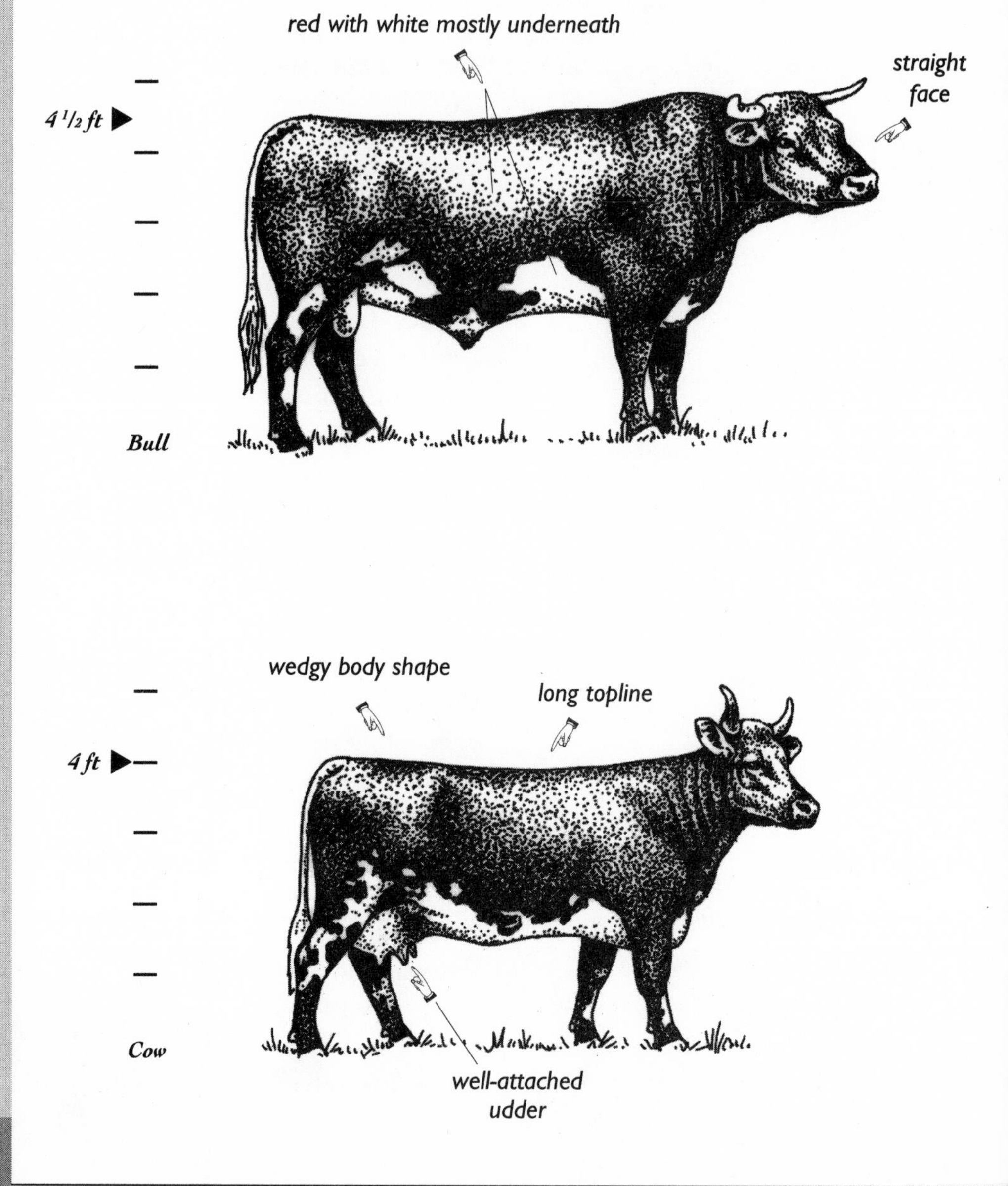

red & white

Norwegian Red

- Synonyms: ***Norsk Rødt og hvitt Fe (NRF)***
- Distribution: ***Limited***
- Registrations: ***N/A***
- Bull's Average Weight: ***2,420 pounds***
- Cow's Average Weight: ***1,320 pounds***

Purpose

Beef and dairy

The cow was originally a dual-purpose breed with a greater emphasis on the dairy traits. The Norwegian Red is said to have decent beef quality, excellent feed conversion, good mothering ability, and plenty of milk.

Origin

Norway

The Norwegian Red was started in 1935, when the Norwegian government decided to consolidate all the country's cattle breeds into one superior dual-purpose breed. The appointed government breeders were able to speed up the usual breeding process by using only the best cows from the large genetic base available to them. The dozen or so breeds present in Norway and incorporated into this new cow one way or another included the Jarlsberg (like the cheese), the Red Tronheim, the Hedmark, the Døle, the South and West Norwegians, the Lyngdal, the Vestland Red Polled, the Vestland Fjord, the Hordaland, the Møre, and the Ramsdal. Topping off this list, some Ayrshire was mixed in along the way. Almost all the breeds used in the breeding process, along with other lesser breeds, have been overtaken by the Norwegian Red. Today, it makes up a commanding 98 percent of the Norwegian cattle population.

Cow Fact:

A maverick is a cow that escaped getting branded.

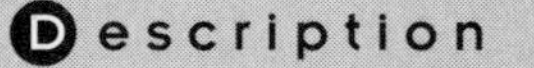

Description

A RED-AND-WHITE COW WITH LOTS OF SMALL, DEFINITE SPOTS.

The cow's reddish brown or mahogany spot color varies from light to almost black. These usually small spots have a more jagged but defined appearance than the spots on other spotted cows. The number of spots range in frequency from a few to covering the whole cow. Other features include a long, straight face and a cool set of long, lyrelike horns. These attributes highlight the Ayrshire's dairy shape: a long, level topline; a well-attached udder; and a spirited step, like the cow knows how great it looks.

red & white with lyre horns

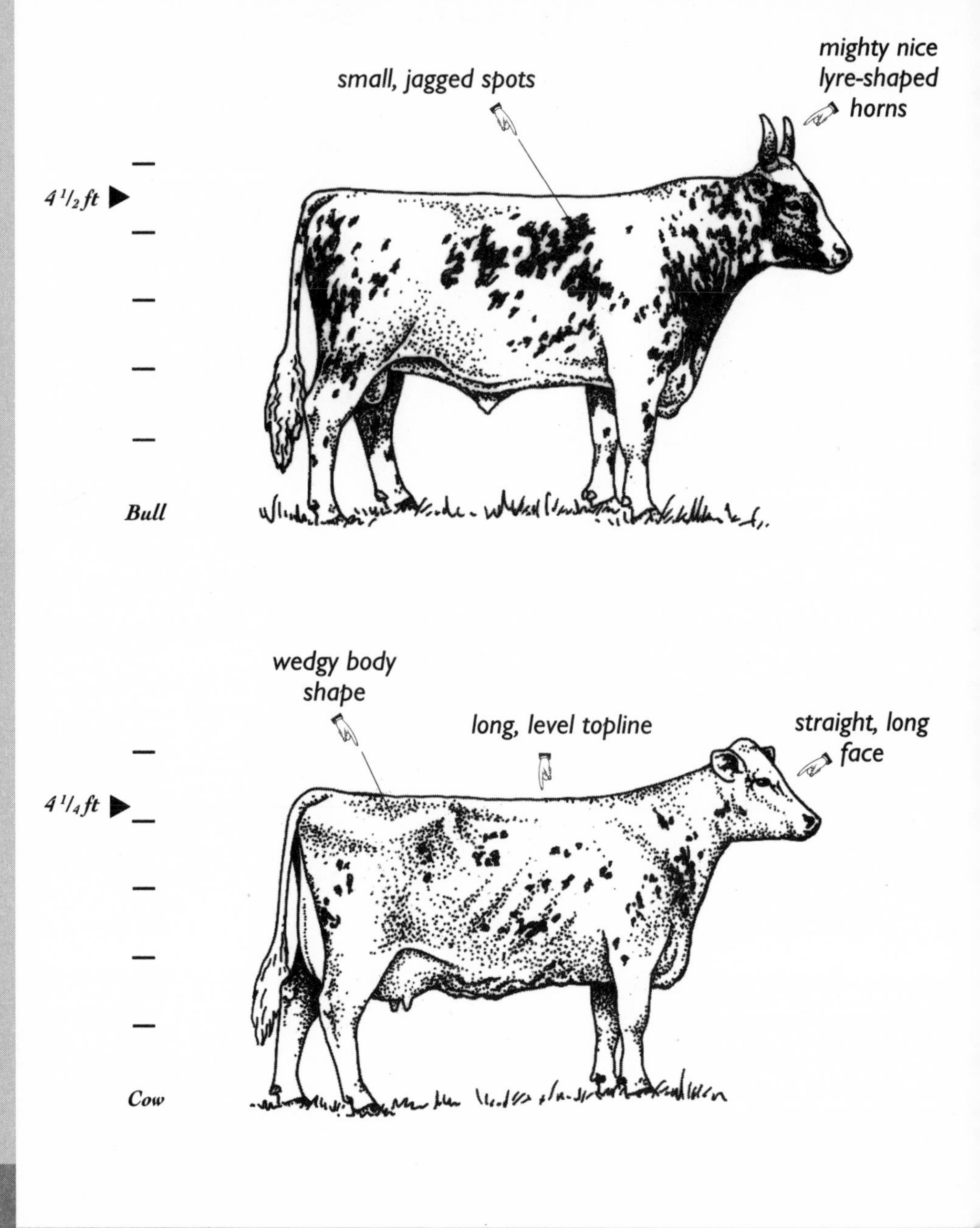

Ayrshire

- Synonyms: ***Cunningham and Dunlop (both obsolete names)***
- Distribution: ***Common***
- Registrations: ***about 19,000***
- Bull's Average Weight: ***1,850 pounds***
- Cow's Average Weight: ***1,200 pounds***

Purpose

Dairy

Farmers consider the Ayrshire a practical breed because of its consistent production of fairly rich milk under almost any conditions. The breed has a higher percentage of cows exceeding the amazing level of 50 metric tons of milk during a milking life than any other breed. Other noteworthy characteristics are excellent grazing ability, adaptability to weather conditions, and acceptable beef value. The top-producing cow was Leete Farms' Betty's Ida with twice-a-day milking for 305 days.

Origin

Ayrshire, Scotland

The Ayrshire started off on the subsistence farms in the southwest Scottish shire of Ayr. The breed's forerunners were probably scraggly scrub and Teeswater cattle, since the practical farmers used any available cows in the region (maybe even the Highland). Later, other imported cows were thrown into the mix, thereby slowly improving the breed's quality. Some of these cows included the early Dutch (pre-Holstein) and the Channel Island cattle. By the end of the 18th century, the Ayrshire's characteristics were close to fixed through the slow process of selective breeding, which completed the transformation from scrub cow to refined, show-stopping dairy cow. The breed was formally recognized in 1814 and the Ayrshire was first imported to Connecticut in the late 19th century.

Guernsey

Description

A SMALL, SPOTTED, BROWN, MILKING COW.

Either brown or fawn, often with white spots of varying sizes on the face and body. The muzzle, hoofs, and udders are unpigmented. A unique characteristic is the yellow tears, actually secretions, that collect around the eyes. In the past, farmers emphasized this trait, believing the color of the tears indicated richer milk. The Guernsey has a refined dairy conformation similar to the Jersey. Both breeds have small, light-colored, dark-tipped horns, but the Guernsey has a less-dished face, a slightly fuller body, and, usually, more spots.

small, brown & white

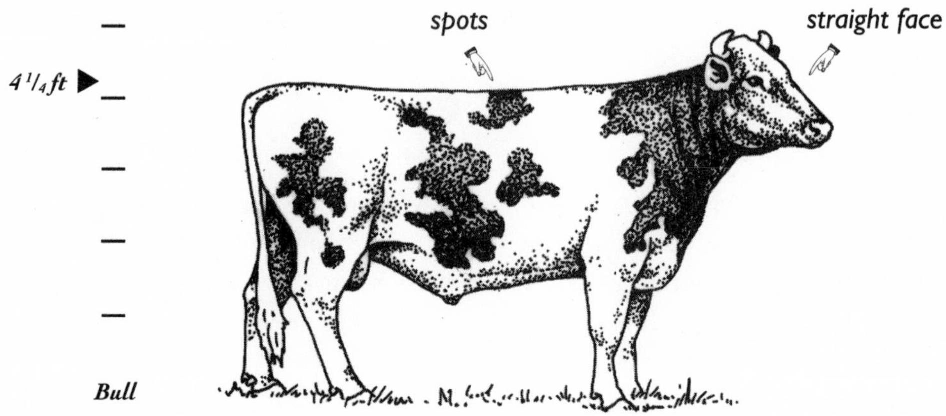

spots

straight face

4 1/4 ft

Bull

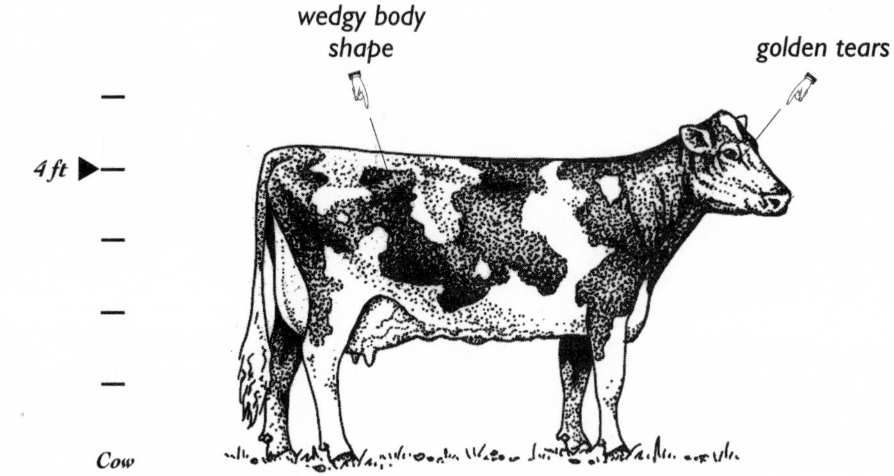

wedgy body shape

golden tears

4 ft

Cow

Guernsey

- **Synonyms:** ***Alderney (obsolete name)***
- **Distribution:** ***Common***
- **Registrations:** ***over 18,000***
- **Bull's Average Weight:** ***1,700 pounds***
- **Cow's Average Weight:** ***1,100 pounds***

Purpose

Dairy

The Guernsey produces the rich and delicious golden Guernsey milk, an immensely popular milk during the 1950s. A few dairies still market it today as an old-fashioned milk; the cream rises. The Guernsey is also noted for decent heat tolerance and a gentle disposition—at least in the cows, if not in the unruly bulls. The cow Fauve Hill Tel Odette gave the record production with twice-a-day milking over 365 days.

Origin

Isle of Guernsey, United Kingdom

A group of monks started the Guernsey breed on the Isle of Guernsey over a thousand years ago. The monks probably imported cattle from the nearby coastal region of France, then home to old breeds like the Froment du Leon and the Norman Brindle. Over the centuries, the Guernsey cows remained in relative isolation, with only a few Jersey cattle making their way across the straits to add some different blood. To keep the breed pure, the government passed a law in the 19th century prohibiting the import of cattle except for slaughter. In 1840, three Alderney cows arrived in New York—there was no distinction made between the Channel Island breeds; all the cattle freighted by the cargo ships were named for the Isle of Alderney (home to the Guernsey, the Jersey, and the now extinct Alderney), the last port of call before the ships crossed the Atlantic. In 1868, American farmers started importing the cows in earnest. The British breed society formed in 1842, the herdbook in 1878.

Cheese Fact:

Wisconsin produces the most cheese in the U.S.A., followed by Minnesota and New York.

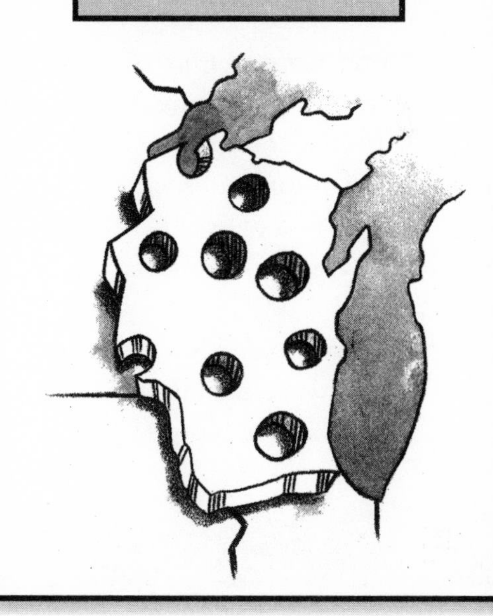

Description

A SMALL, BROWN, DOE-EYED COW.

A brownish shade of fawn is the usual color. Other colors, such as mulberry, gray, and spotted white, occasionally occur. Whatever the color, it blends to a lighter shade underneath, and a light band appears around the muzzle. The favored muzzle color is black, but a buff-colored muzzle is allowed. Other features include a broad, dished face; prominent eyes; black skin pigment; a small chest; and small, incurving horns.

The Jersey is noted for a refined, cute appearance and for its angularity, as seen in the protruding hips.

brown

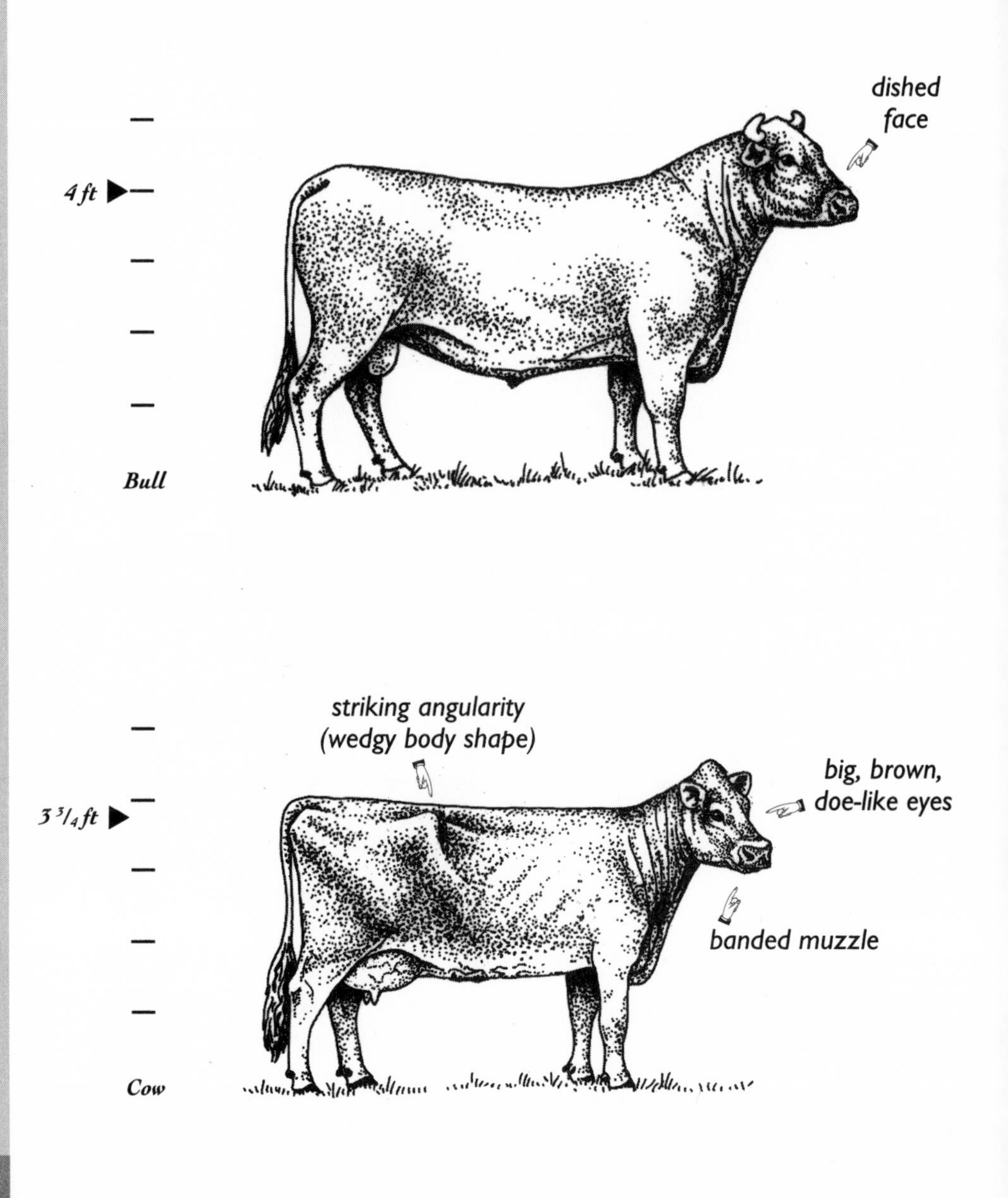

Jersey

- Synonyms: ***Jersias (French) and Alderney (obsolete name)***
- Distribution: ***Ubiquitous***
- Registrations: ***over 60,000. Found mostly in the South.***
- Bull's Average Weight: ***1,500 pounds***
- Cow's Average Weight: ***1,000 pounds***

Purpose

Dairy

Of all the breeds, the Jersey yields the richest milk in percentage of butterfat and protein. And yet, the amount of milk the cow produces is equivalent—for its body weight (it produces over 12 times its own weight)—to what the Holstein produces, without the intensive labor. The cows have a gentle nature, early maturity, and intelligence; in addition, they are the most heat tolerant of the dairy cows. These traits highly recommend the breed as a family cow. The cow Rockhill Favorite Deb produced the record milk yield with twice-a-day milking over a year.

Origin

Isle of Jersey, United Kingdom

Researchers have yet to pinpoint the Jersey's ancestry. Conflicting theories claim that it originated from the cattle of nearby Brittany and Normandy and from cattle as far away as the Zebu on the Indian subcontinent. Whatever the source, three things can be said about the Jersey's history: it is guaranteed purebred since 1763, at which time the government banned the import of other milk cows; it is recorded as purebred for the last 600 years; and it has resided on the Isle of Jersey for at least 1,000 years. Surprisingly, the Isle of Jersey has a total cattle population of only a few thousand head. To enable this number to find enough forage on the limited pastures, the farmers worked out a special grazing system: the cows are tethered by their horns for a couple of hours at a time to graze in a small circle. A breed standard (similar to a herdbook) formed in 1844; the U.S. breed society formed in 1868.

Brown Swiss

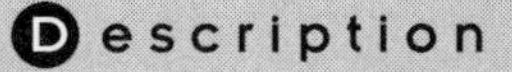

Description

A BIG BROWN COW.

The cow's color may vary from a light to a dark shade of brown, but it is always lighter underneath. The muzzle is banded. The coloration resembles the Jersey's, though the larger Brown Swiss presents a fuller body shape and a more rugged appearance. Other features include large fuzzy ears, a black switch, a straight face, a stocky build, and short, dark-tipped horns.

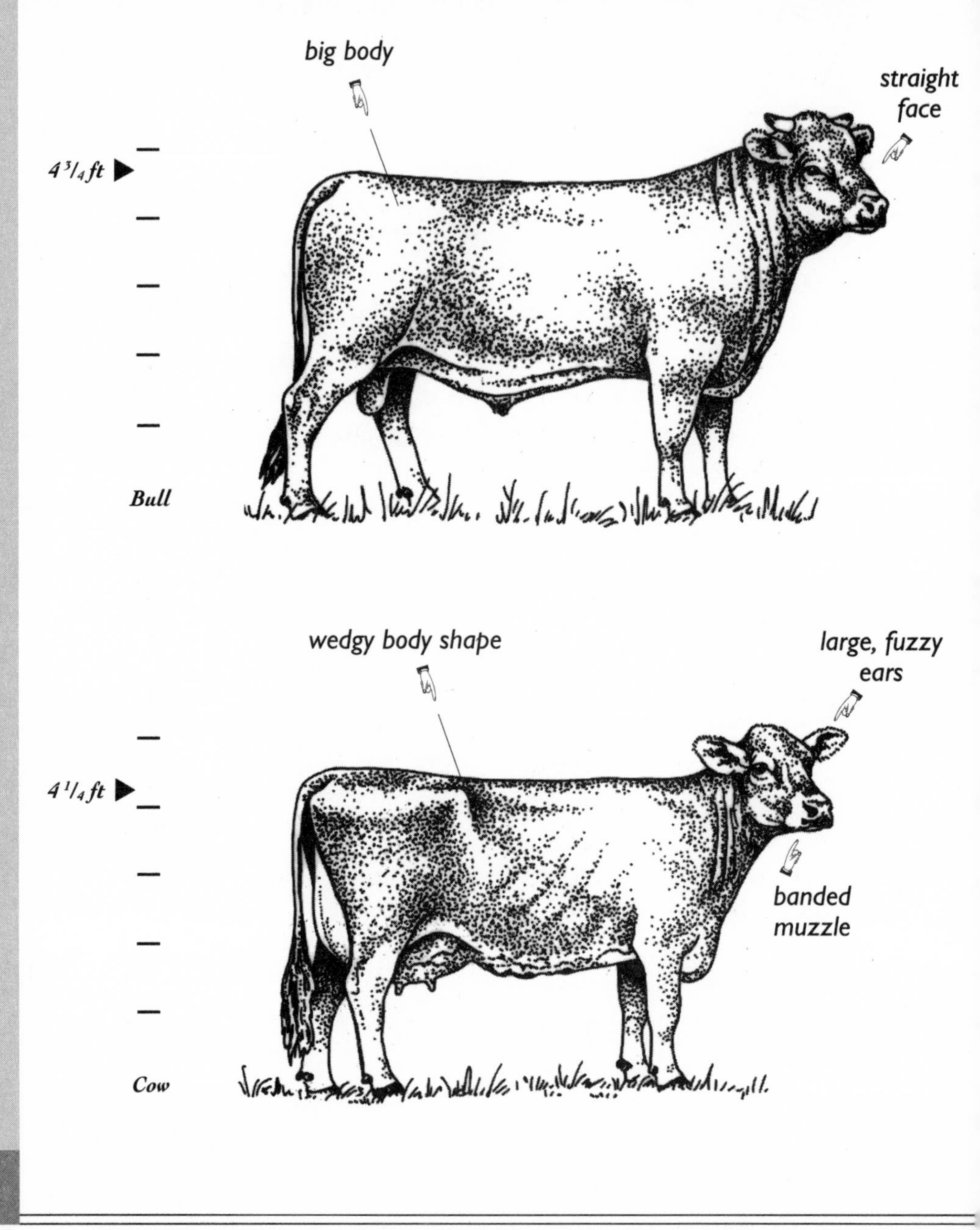

Brown Swiss

- Synonyms: ***American Brown Swiss (ABS), Braunvieh, and Schweitz***
- Distribution: ***Common***
- Registrations: ***over 12,000***
- Bull's Average Weight: ***2,000 pounds***
- Cow's Average Weight: ***1,400 pounds***

Purpose

Dairy

The Brown Swiss was originally a triple-purpose breed; farmers first shifted its focus to beef and then to dairy production. Today in the U.S., the Brown Swiss is a singular dairy cow. It is second only to the Holstein in total milk yield, and the Brown Swiss far exceeds the Holstein in total milk-solid content. In 1980, the cow L&S Telstar Sally May Twin gave the record Brown Swiss milk yield and the highest butterfat percentage of the dairy breeds. The total butterfat weight works out to a whopping 2,067 pounds. Other noted qualities of the Brown Swiss are a strong constitution, decent beef quality, and a docile, unexcitable nature.

Origin

Central Switzerland

According to those archaeologists who carbon-date fossilized cow bones, the Brown Swiss might be one of the oldest cow breeds. The cow only became known outside of Switzerland in the mid-19th century. At that time, a flourishing market in France and Italy favored the beef produced from the Swiss breed and made it popular. To help meet the demand, Swiss farmers formed a nationwide cooperative which increased the breed's numbers and spread the cow's range throughout the country. Later, in America, the Brown Swiss changed in conformation to the present-day refined dairy shape. The change in the American cow was big enough that it is now considered a separate line from the Swiss cow, which is called Swiss Brown or Braunvieh.

Cow Fact:

Cambridge, Massachusetts started off as a cow pasture and Boston was actually laid out by cow paths.

Description

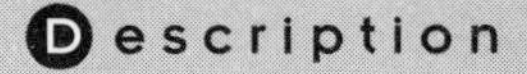

A BIG YELLOW COW.

The classic color is yellow, but variations include a creamy color, reddish yellow, and black. The cow's light-colored skin highlights the eyes and the udder and is most noticeable on the darker-colored cows. Other features consist of a Roman-nosed face (i.e., it's long and straight), a hairy foretop, and short horns. The cow is noted for its big, beefy conformation: a long, level topline and strong, muscular hindquarters.

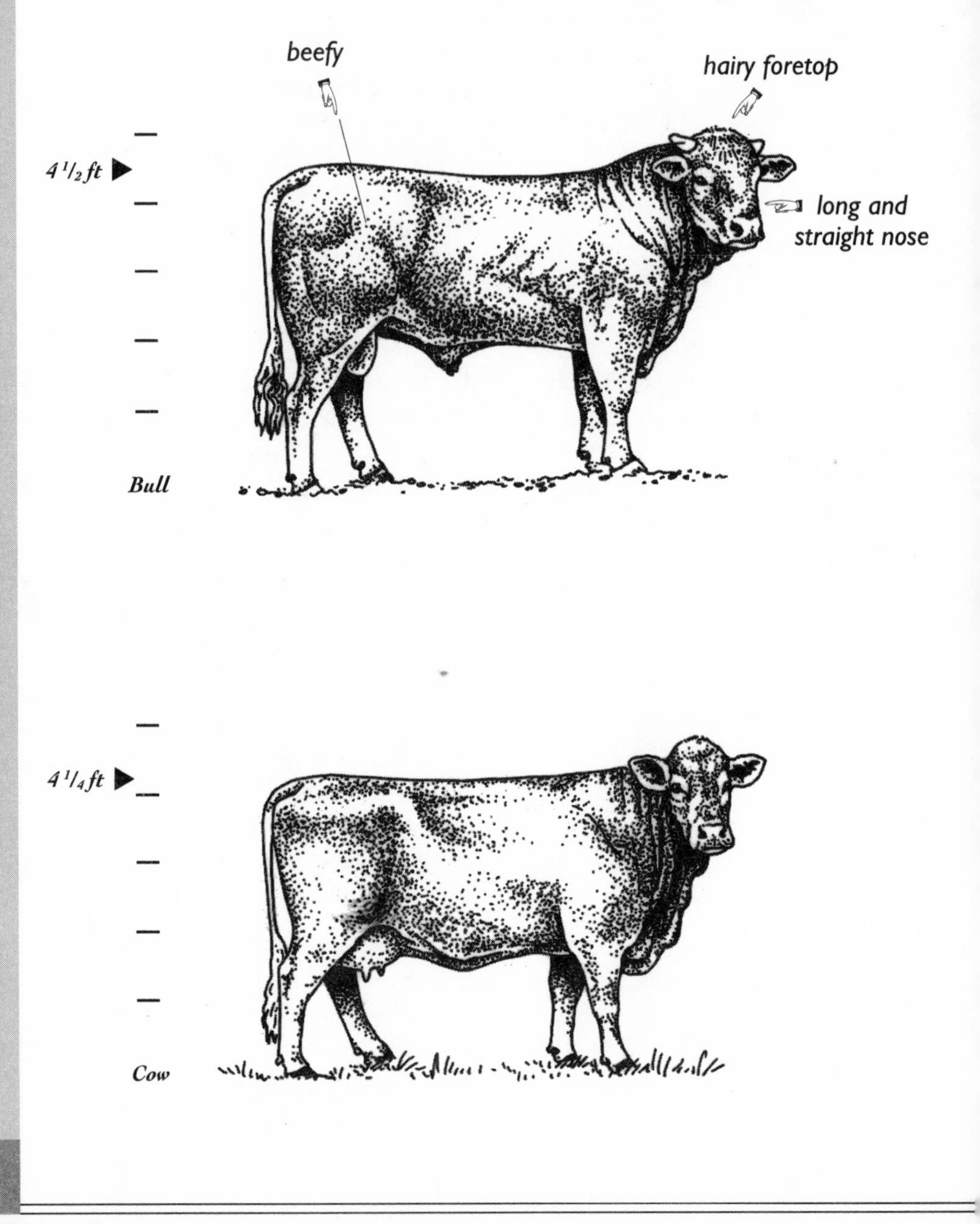

yellow

Gelbvieh

- Synonyms: ***German Yellow and Einafarbig Gelbes Hohenvieh***
- Distribution: ***Common***
- Registrations: ***about 21,500***
- Bull's Average Weight: ***2,500 pounds***
- Cow's Average Weight: ***1,600 pounds***

Purpose

Beef and dairy

European farmers have separated the breed into two different lines, one for beef production and one for dairy purposes. American farmers mostly raise the Gelbvieh for beef, even though it possesses some dairy traits such as decent milk yield and good maternal care. The Gelbvieh is noted for fast growth, a quality carcass, feedlot efficiency, and an even temperament.

Origin

Bavaria, Germany

During the Gelbvieh's development, the family tree became a little complicated. The cow arose from four main German triple-purpose breeds: the Glan-Donnersburg, the Yellow-Franconian, the Limpurg, and the Lahn. Other breeds used to some degree in the confluence were the Swiss Brown, the Bernse, and the Simmental. This hodgepodge of breeding activity started in 1856, but it was 1897 before the cows were uniform enough to be called a breed and to have a breed society. Later, in the 1920s, the Gelbvieh's beef conformation was finally fully set, giving us today's shape. The Gelbvieh was imported to the United States in 1972, during the Euro-cow movement. "Gelbvieh" is pronounced "gelp-fee"; in German it means yellow cow.

Cow Fact:

A cow's heart beats between 60 and 70 beats per minute.

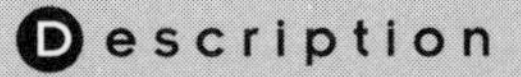

Description

A BIG BLOND COW.

The color can range from almost white to dark brown, and a lighter shade may occur around the eyes and undercarriage. Physical characteristics include short, smooth hair (which shows off the muscularity); a slightly dished face; short, curving horns; a long, straight topline; and black hoofs. The breed is similar to the Limousin but larger and more muscular. The Blonde is one of the largest and heaviest of the French breeds.

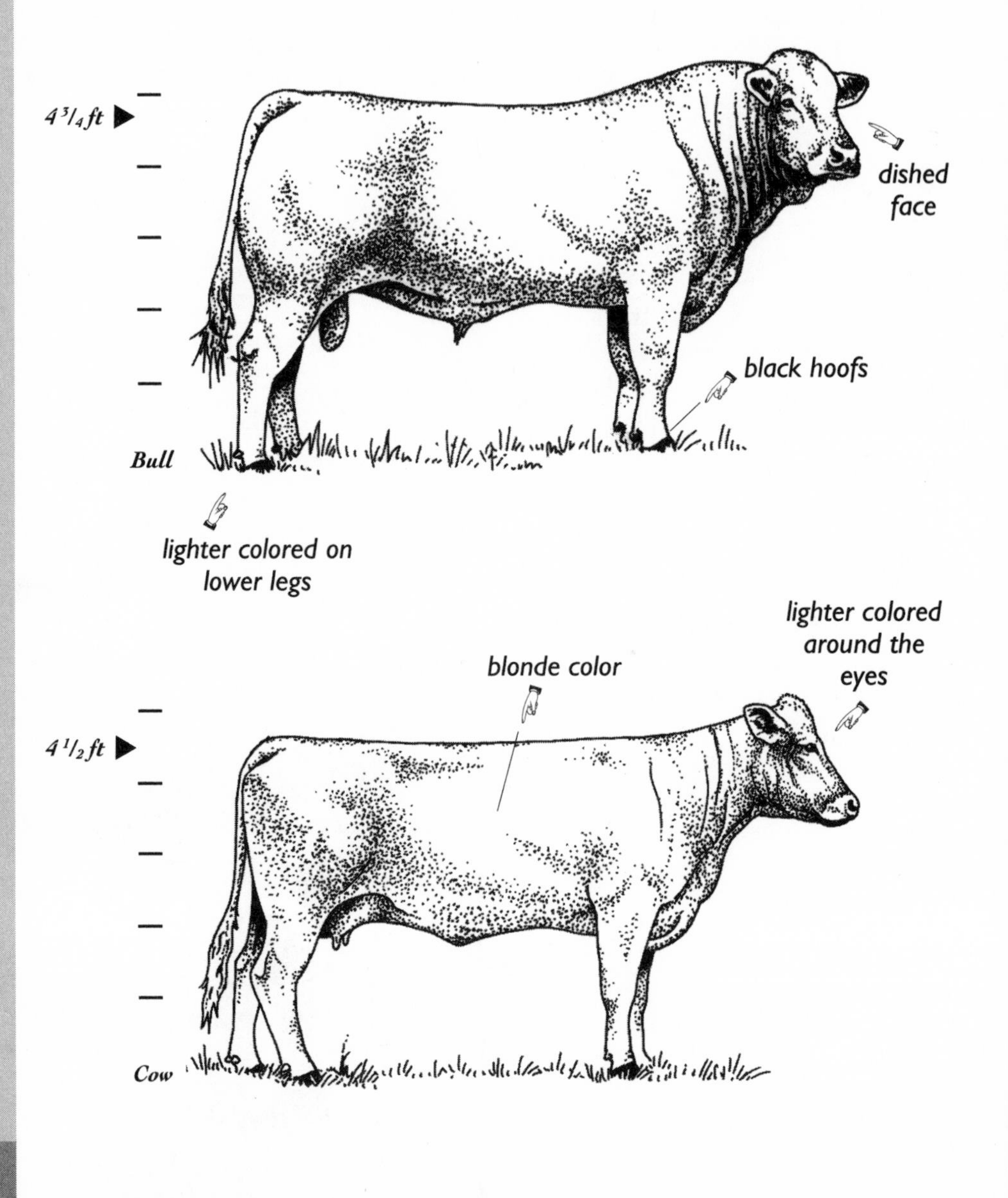

blond

Blonde d'Aquitaine

- Synonyms: ***Aquitaine Blond and Garrone***
- Distribution: ***Sparse***
- Registrations: ***about 1,000***
- Bull's Average Weight: ***2,640 pounds***
- Cow's Average Weight: ***1,760 pounds***

Purpose

Beef

Now a single-purpose beef cattle, the Blonde d'Aquitaine worked as a draft animal before the age of tractors. A few of its characteristics are a docile disposition, strong limbs (from the draft heritage), decent milk yields, calving ease, and heat tolerance. The breed also has good insect tolerance, partly because of its ability to flick bugs off with its skin. This twitching action of the skin resembles the skin movement of a horse, and among cattle it is unique to the Blonde; other cow breeds are left with only their tails to chase away flies.

Origin

Garrone River region, France

The Blonde d'Aquitaine was developed in the early 1960s from three lines native to the southwest of France: the Pyrenees, the Garronais, and the Quercy. The Pyrenees was used mostly as a dairy cow; the other two were closely related beef and draft cattle. These two related cow lines once had nearly identical names: Garronais du Plaine and Garronais du Coteau. They diverged when the Garronais de Coteau was crossed with the Limousin, creating the Quercy. History dates the original lines to Roman times, at least. The Blonde d'Aquitaine was imported to the United States in 1972 and a U.S. breed society formed the next year.

Cheese Fact: *Limburger cheese takes 2 months to develop that staggering smell. This wondrous odor is caused by yeast and* bactrium linens *decomposing milk protein.*

Description

A LONG, SLEEK COW.

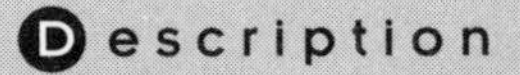

The color varies from wheat to rust red. On the darker-colored cows, the hairs contrast against the lighter shading underneath, around the eyes, on the switch tip, and on the nose. Other traits include a relatively short face; a broad, high forehead; a short neck; short to medium-length horns; and decent muscling, especially in the hindquarters. The conformation is roughly similar to the lighter-colored Blonde d'Aquitaine and Gelbvieh.

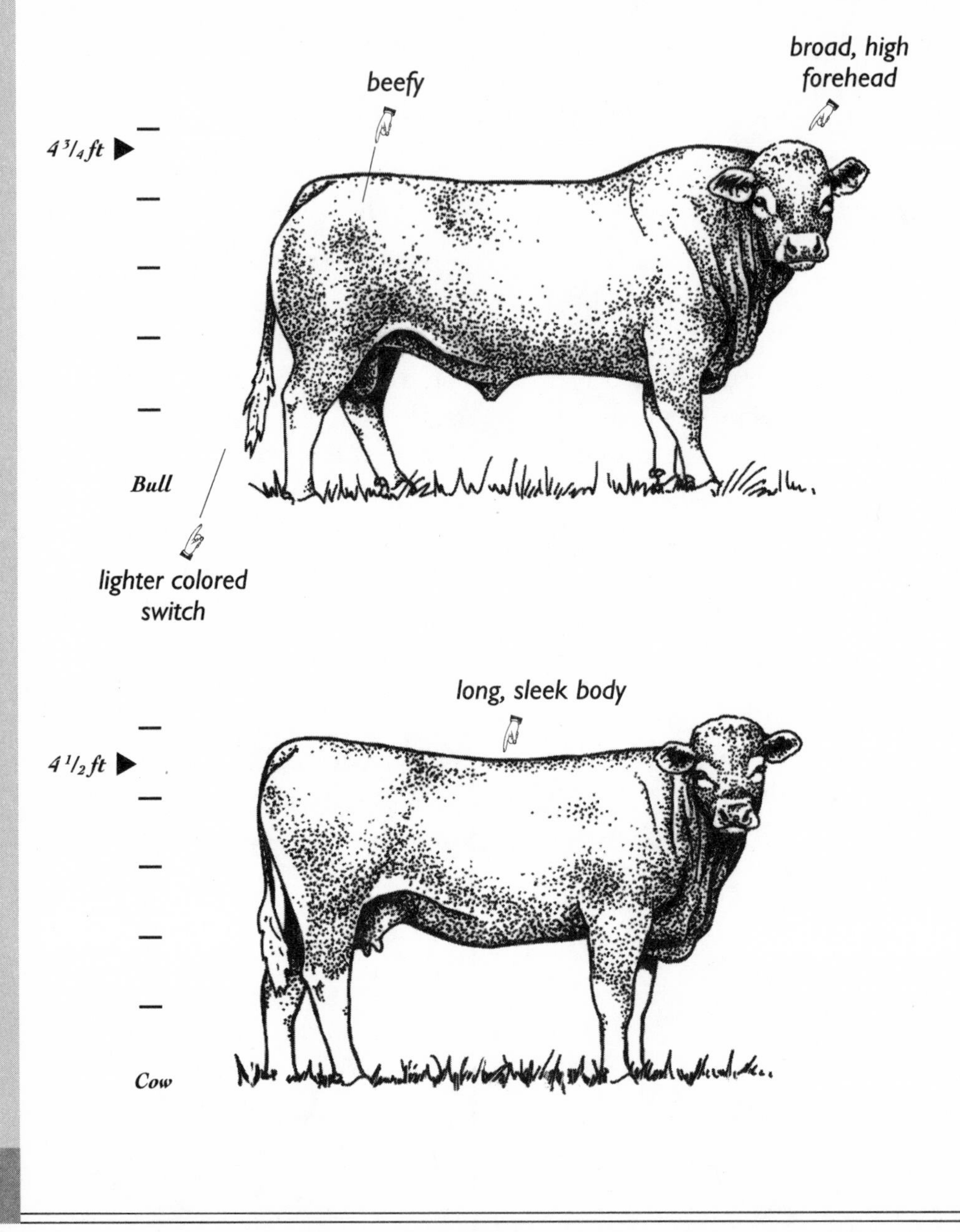

brown

Limousin

- Synonyms: ***None***
- Distribution: ***Ubiquitous***
- Registrations: ***about 59,000 (Raised mostly on Cadillac ranches.)***
- Bull's Average Weight: ***2,350 pounds***
- Cow's Average Weight: ***1,650 pounds***

Purpose

Beef

The Limousin was formerly a draft animal noted for strength and speed. Today, its high-quality beef, good cutability, and nicely developed hindquarters (i.e., money cuts) make it noted as a modern beef type. Other traits include manageability, feeding efficiency, and calving ease because of a small head. Criticized for being a bit unreasonable, the breed does well out on the open range, where it needs its tough temperament and increased awareness.

Origin

Limoges, France

The Limousin originated in central France, near the town of Limoges and its namesake mountain range. Twice, in the late 18th and mid-19th centuries, farmers tried cross-breeding the Limousin to increase its overall size, but both times the effort was abandoned because the region's limited resources could not accommodate the beefed-up animals. At the end of the 19th century, Charles de Leobarg worked to improve the animals using natural selection. Farmers of the 1900s further emphasized the cow's conformation, notably the deep chest, the strong topline, and the beefy hind-quarters. The famous Lascaux cave cow drawings, which archaeologists say date from 25,000 years ago, share the Limousin's range and, some say, the Limousin's conformation. Today's cow still sort of has that wild-eyed look.

The Limousin also has this old French folk song by Pierre Dupont written about it (author's translation):

Les voyez-vous les belles betes,
Creuser profond et tracer droit.
Bravant la pluie et les tempetes,
Qu'il fasse chaud, qu'il fasse froid.

You see them, the cool cows,
Digging deep and pulling bold.
Braving rain and lows,
When it's hot, when it's cold.

Description

A FINCHBACKED, BLACK-NOSED COW.

The color ranges from a dark yellow to a light red and often features a finchback—a ridge of light-colored hairs along the back. Other features include black-pigmented skin; short, black-tipped horns; and a slender, dual-purpose conformation. The dairy part of the conformation is readily apparent on the cows, whose udders are bigger than those of most other beef cattle. The Tarentaise is smaller than the other French breeds, but is about the same size as most of the British breeds.

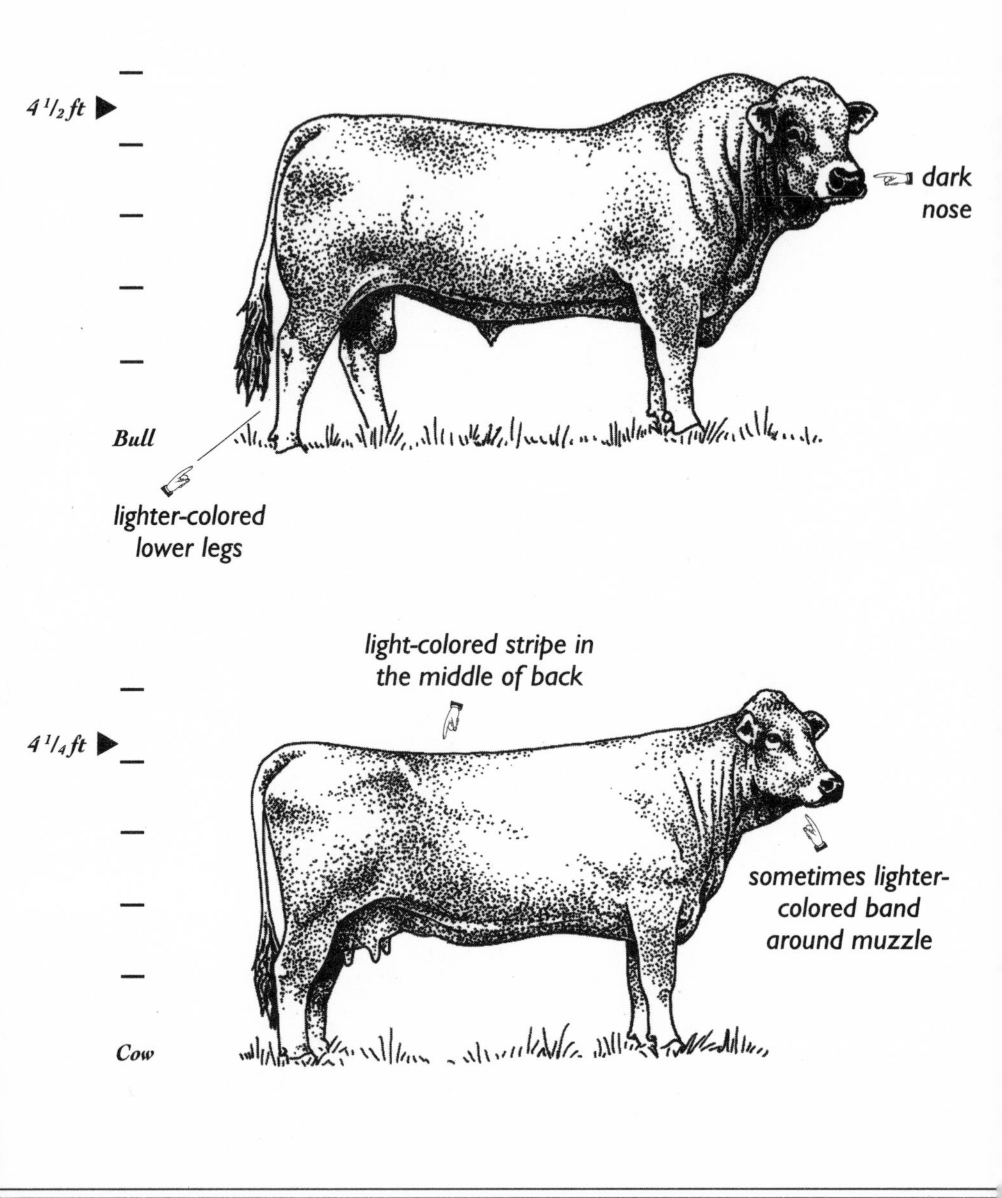

brown

Tarentaise

- Synonyms: ***Tarina (Italian)***
- Distribution: ***Limited***
- Registrations: ***about 2,000***
- Bull's Average Weight: ***2,100 pounds***
- Cow's Average Weight: ***1,200 pounds***

Purpose

Beef and dairy

In the past, the Tarentaise was a dual-purpose breed with greater emphasis on the dairy traits; in France this is probably still true. American farmers raise it solely for beef production. The Tarentaise has a divergent genetic background, which gives a good hybrid vigor when the cow is crossed with English breeds. The cattle are hardy and accommodating, though some criticize them for being a touch excitable.

Origin

Haute Savoy, France

The Tarentaise is a descendant of the native cattle found around the Isere River in eastern France. Over the years, the cow's only development involved the mixing of the larger valley cattle and the smaller mountain cattle. The farmers placed an emphasis on dairy traits. The region's tough topography and climate also influenced the cows, shaping their muscles and making them hardy. Early in the 20th century, ranchers first exported this adaptable breed to rugged North Africa. The Tarentaise was recognized in 1859, a herdbook formed in 1880, and the cow was imported to the United States in 1973.

Cow Fact:

A 1,000 pound cow produces on average 10 tons of manure a year.

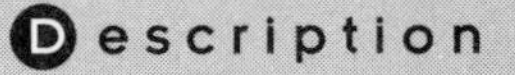

Description

A BIG, BURLY, CHERRY RED COW.

In addition to its cherry red color, the cow has a white switch, a rose-colored nose, black hoofs, and, occasionally, white spots on the udder. A separate black-bodied Salers line also exists. Other features include medium-length, black-tipped, lyre-shaped horns; a large, triangular head; and an overall big, burly conformation. The Salers is the darkest of the French cows; it is closer in color to the red English breeds.

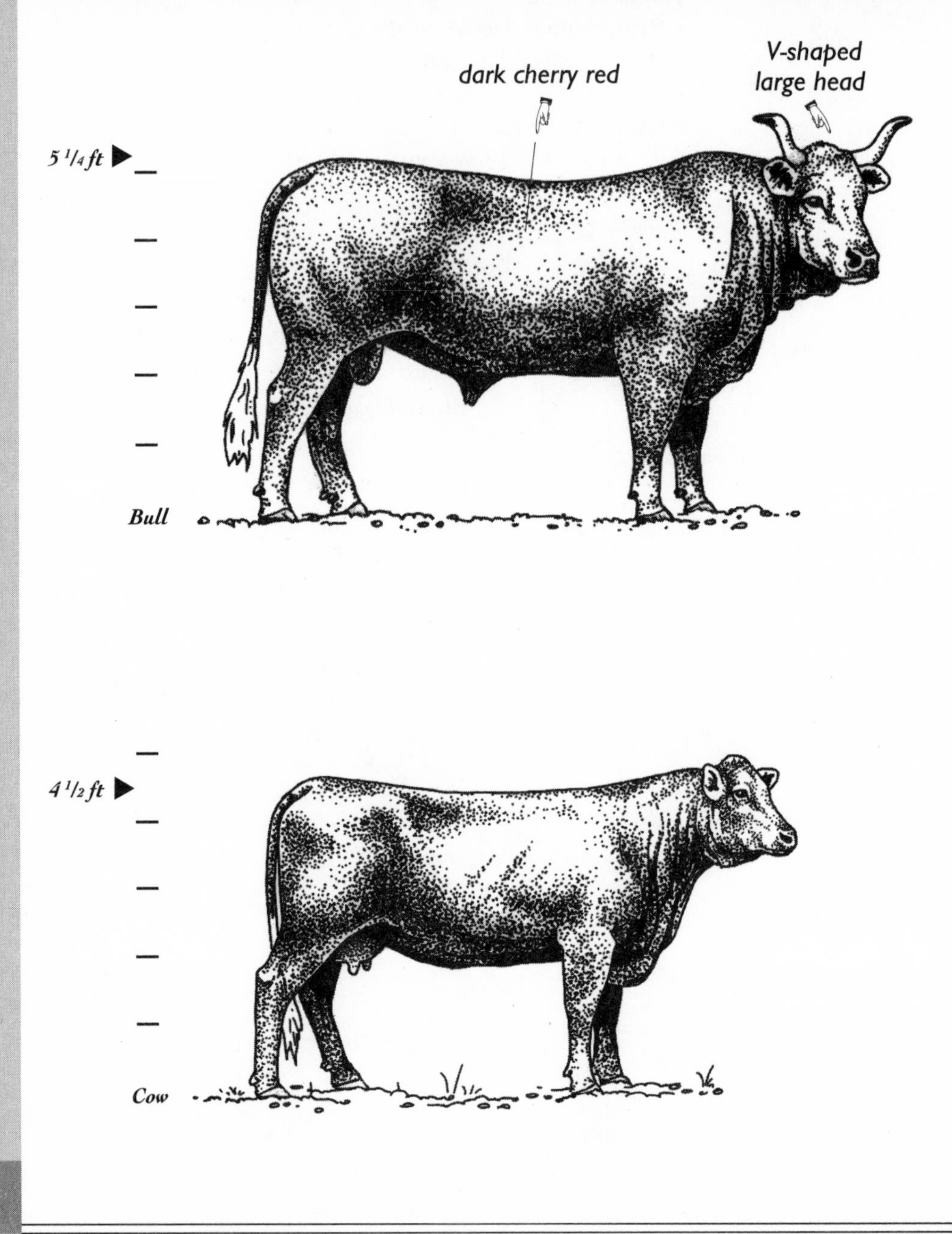

Salers

- **Synonyms:** *Auvergnate (French)*
- **Distribution:** *Common*
- **Registrations:** *about 23,000*
- **Bull's Average Weight:** *2,500 pounds*
- **Cow's Average Weight:** *1,500 pounds*

Purpose

Beef and dairy

The Salers was first developed as a dual-purpose breed, though it was also put into the yoke for plow pulling. Today, the breed is predominantly raised on beef ranches, where it often functions as a maternal cow (that is, it gets to raise calves year in and year out). Besides imparting decent beef quality to its offspring, the cow calves easily and has an ample milk supply.

Origin

South-central France

The breed originated in south-central France and is named for a small Alpine town in the heart of the Auvergne region. In this place, the Salers began as a one-year cow: farmers needed the cows to head for the lush spring pastures with new calves every year. The cows arrived at the pastures at peak season, almost guaranteeing a surplus of milk. This meant the calves would get plenty fat, and any excess milk could go into making some fine, smelly cheese. Besides the emphasis on consistent yearly production, little other development took place until the early 19th century, when breeders imported some Devon cows to increase the Salers' body size and milk production. The Salers later returned the favor to the Devon when it needed some improvements of its own. Official recognition for the Salers occurred in 1853. In 1990, it was named the official U.S. Olympic team beef. To pronounce the cow's name the French way, say "*say-lair*."

Cow Fact:

In an average herd, there is one bull to every 30 cows.

South Devon

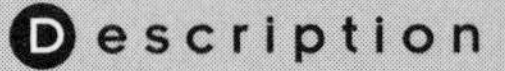

Description

A BIG, RED, MELLOW COW.

The usual coloration is a light brownish red with the tip of the switch a bit lighter. Complementing the soft, curly hair, the loose hide has a light red pigment. The South Devon is known as a gentle cow with a generally plain look emphasized by a wide muzzle and dark, deep eyes. Even the medium-length horns don't intimidate anyone too much, since, on the bulls, they curve downward. The South Devon is among the largest of the British breeds and has a lighter color than either the Devon or the Sussex.

all red

5 ft

Bull

big, light-colored body (sandy red)

4 1/2 ft

buff, wide muzzle

Cow

South Devon

- Synonyms: ***South Hams (obsolete name)***
- Distribution: ***Limited***
- Registrations: ***about 1,800***
- Bull's Average Weight: ***2,400 pounds***
- Cow's Average Weight: ***1,450 pounds***

Purpose

Beef and dairy

Traditionally a dual-purpose breed, the South Devon today also functions as a draft oxen on the small farms where it's often raised. Some of the breed's noted qualities are a docile temperament, a long life, and heat and insect resistance. The South Devon's dairy qualities are also quite decent; in fact, the breed is somewhat famous for its milk, which can be used to make the renowned Devonshire clotted cream.

Origin

South Hams, England

The South Devon's history on the small farms of South Hams (just south of Devon) goes back nearly 400 years. The exact makeup of the breed is a little clouded: some hold that the Devon, the Guernsey, the Normande, the Swiss Brown, the Gelbvieh, and the Blonde d'Aquitaine all had some genetic input. Probably the truest connections are with the Devon, its northern neighbor, and the Guernsey, its neighbor off the coast. The Devon link shows itself in an overall beefiness and the red color, while the Guernsey link includes similar blood type and comparably rich milk, high in protein and butterfat content. The English South Devon breed society formed in 1891. Henry Wallace, the former U.S. vice president, was the first to import the breed to the United States. In the 1930s and 40s the cattle were imported in greater numbers, and a U.S. breed society formed in 1974.

Cow Fact: *A cow stands up and sits down about 14 times a day.*

Description

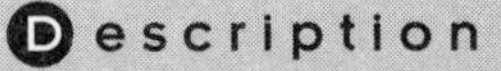

A RECTANGULAR, RUBY RED COW.

The thick, red-colored hair is slick; it dapples with age. Under the hair, the orange-tinted skin is visible around the eyes and muzzle. The medium-length horns are creamy white and have black tips. The Devon has a good beef conformation: nicely muscled and stocky. The even rarer Milking Devon (the other Devon line) shows the greater angularity and refinement of traditional dairy conformation. The Devon is cleaner cut and more handsome than either of the other nearby British breeds, the big-shouldered Sussex and the lighter-colored South Devon.

red

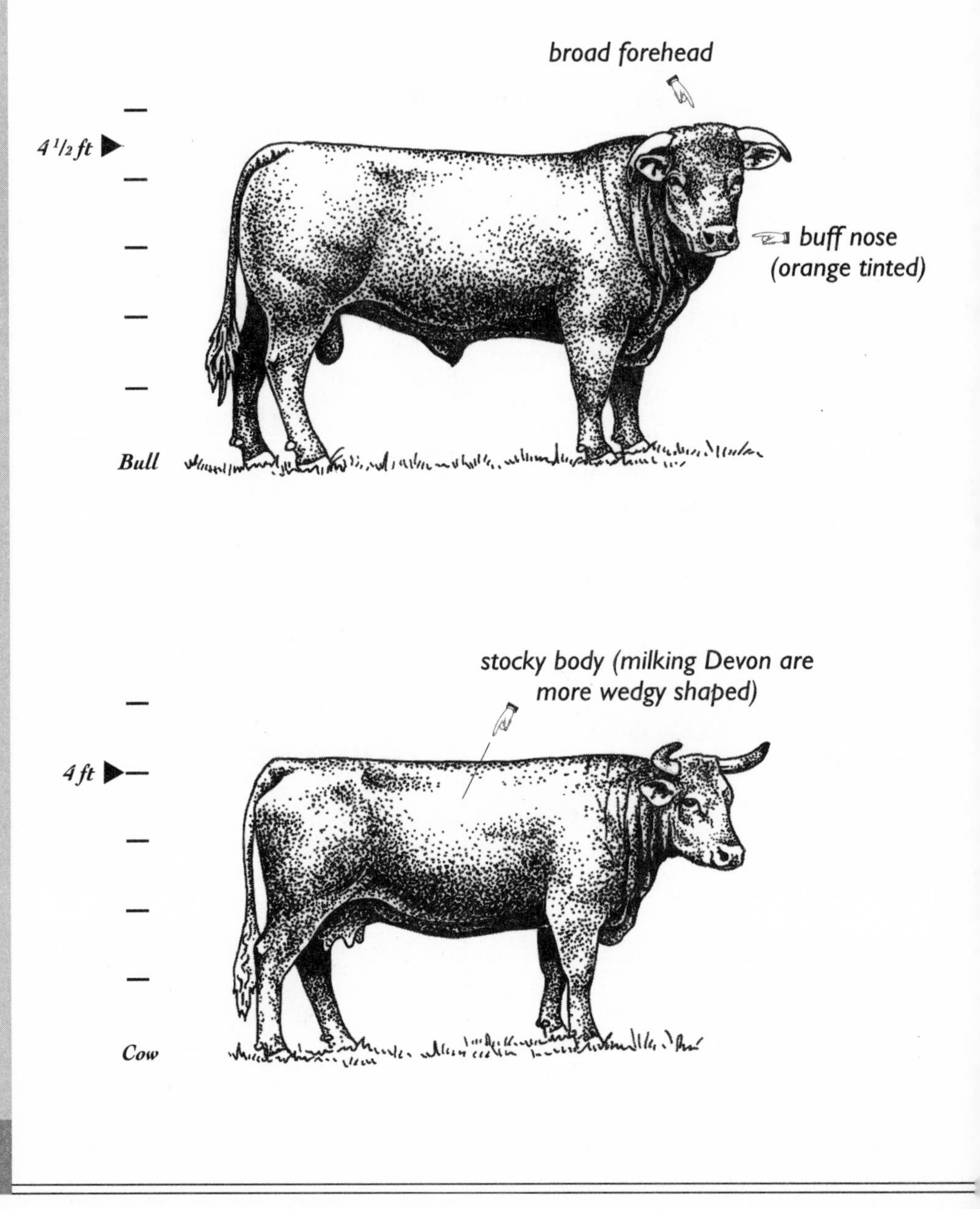

Devon

- Synonyms: ***Beef Devon, North Devon, Red Devon, and Red Ruby***
- Distribution: ***Critical***
- Registrations: ***about 170***
- Bull's Average Weight: ***2,200 pounds***
- Cow's Average Weight: ***1,100 pounds***

Purpose

Beef, dairy, and draft

The breed is separated into two lines: the single-purpose beef Devon and the triple-purpose Milking Devon. The beef Devon is known for early maturity, a fine-boned frame, and tender cuts of beef, while the Milking Devon is not exclusively a milker but a do-it-all cow: milk, beef, and draft. Both lines are noted for their adaptability, economic growth, and cold-weather tolerance. Breeders have thought so highly of the breed's hot-weather tolerance that they have used the Devon as a foundation for some new tropical breeds.

Origin

Somerset and Devon Counties, England

The pilgrims first brought the Milking Devon to North America from their farms around Exmoor, north of Devon. Descendants of these cattle continue to be purebred, making the Milking Devon line the oldest continuously bred cattle breed in the U.S. Back in the colonial days, almost all the cows were like the Milking Devon—all-purpose animals supplying not only milk, but also leather, animal traction, cheese, and at the end of their long life, beef. The old line is still raised on a historical Vermont farm, the place of its American sanctuary throughout the years. In the 1950s, the breed diverged into two different lines when breeders wanted a more industrial steer for commercial purposes. To get that beefier conformation, they crossed in some Salers' blood. A U.S. herdbook formed in 1855.

Cow Fact:

An average cow with 2 milkings produces about 10 gallons of milk a day, or 50 to 60 pounds.

I lost 700 lbs in two weeks!!
LOSE WEIGHT NOW!!
THE FARMER TED WAY
CALL 555-MILK

Description

A BIG, RED, RUGGED COW.

The color is usually some shade of dark red, or, as some say, almost a reddish shade of black. The coat itself is sleek and short, though it turns long and curly in the winter. The dark hairs contrast sharply with a light skin, notably on the muzzle and around the eyes. The horns are medium length. The Sussex's conformation is similar to the Devon's, except that the Sussex is a bit rougher and more muscled in the shoulders.

all red

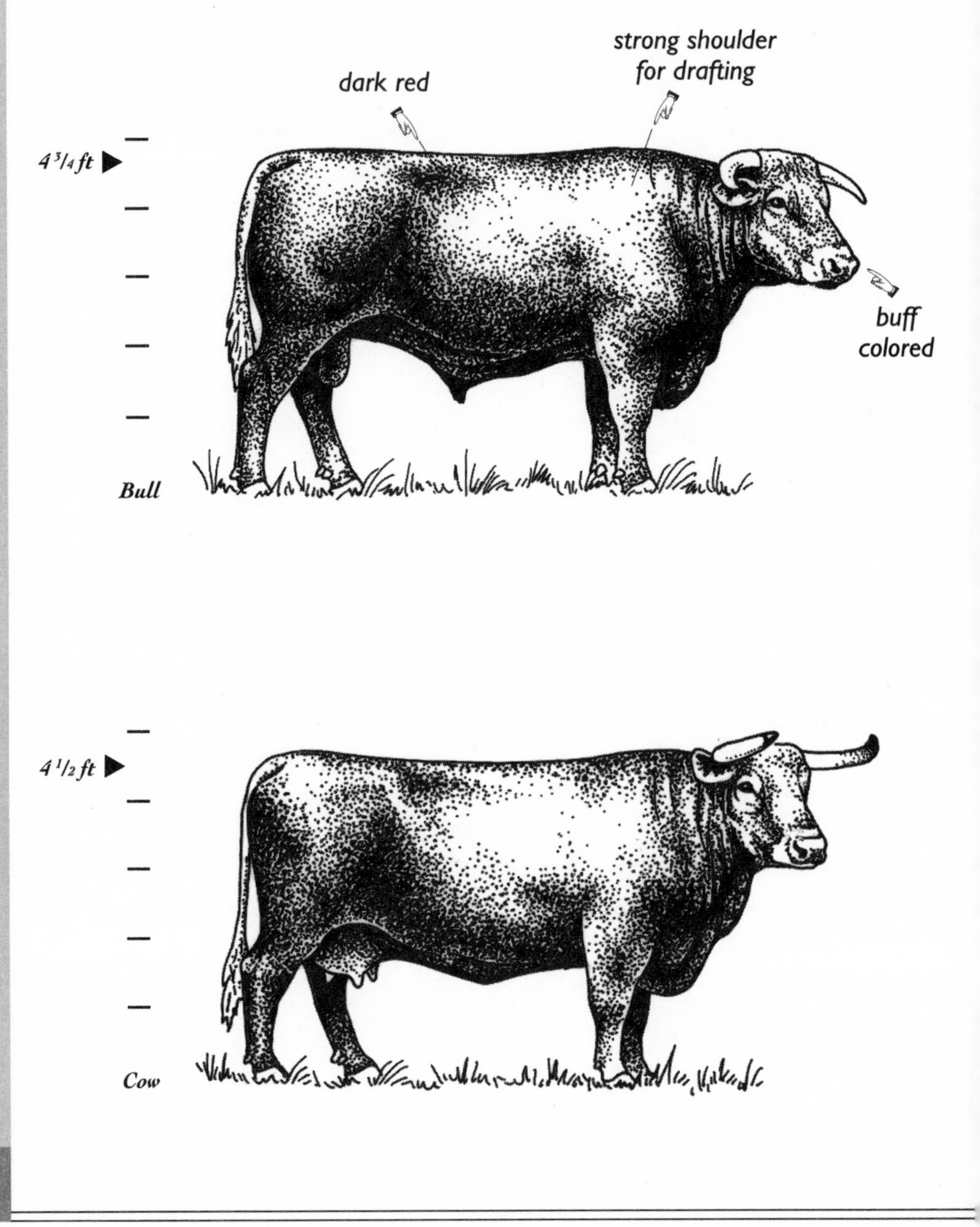

Sussex

- **Synonyms:** *None*
- **Distribution:** *Limited*
- **Registrations:** *N/A*
- **Bull's Average Weight:** *2,100 pounds*
- **Cow's Average Weight:** *1,300 pounds*

Purpose

Beef

In the old days, the Sussex pulled plows, providing beef only at the end of its working life. Its thrifty grazing habits, particularly in hot weather, are among its more noted traits. The cow's tolerance of heat arises partly from having twice the normal number of sweat glands (an adaptation also found in a native hot-weather breed, the Africander) and partly from its general hardiness. Other traits include early maturity, quality beef cuts, resistance to biting insects, a calm disposition, and sound legs.

Origin

Sussex and Kent Counties, England

The Sussex, like the Devon and the Hereford, descended from the native red cattle of southern England. Around the time of the stormin' Norman conquest (1066), the Sussex was developed into a strong, muscular draft ox. Farmers needed and bred big-shouldered oxen capable of plowing the heavy clay soils found in Kent and Sussex Counties. This adaptation was the main distinction between the Sussex and the neighboring big red breeds. The Sussex did its previous job so well that only in the last 50 years or so did it need any refinement; this entailed beefing up the hindquarters.

One of the Sussex's moments of true glory came when Rudyard Kipling immortalized it in the poem "Alnascher and the Oxen." Briefly, the poem is about a poor farmer who has been around his cattle too long, in the far-off fields of Sussex County.

"To a luscious sound of tearing, where the clovered herbage rips,
Level-backed and level-bellied watch 'em move.
See those shoulders, guess that heart-girth,
Praise those loins, admire those hips . . ."

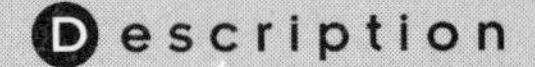

Description

A POLLED, RED COW.

The color may vary from light to dark red; occasional touches of white may show up on the udder and the switch tip. The buff-colored skin is visible in several places: on the muzzle, on the teats, and around the eyes. The Red Poll, as its name proudly advertises, is and always has been naturally polled. Other traits include a straight topline, a small forehead, or poll, and a decent dual-purpose body shape. The Red Poll is said to be a good-looking, muscular cow.

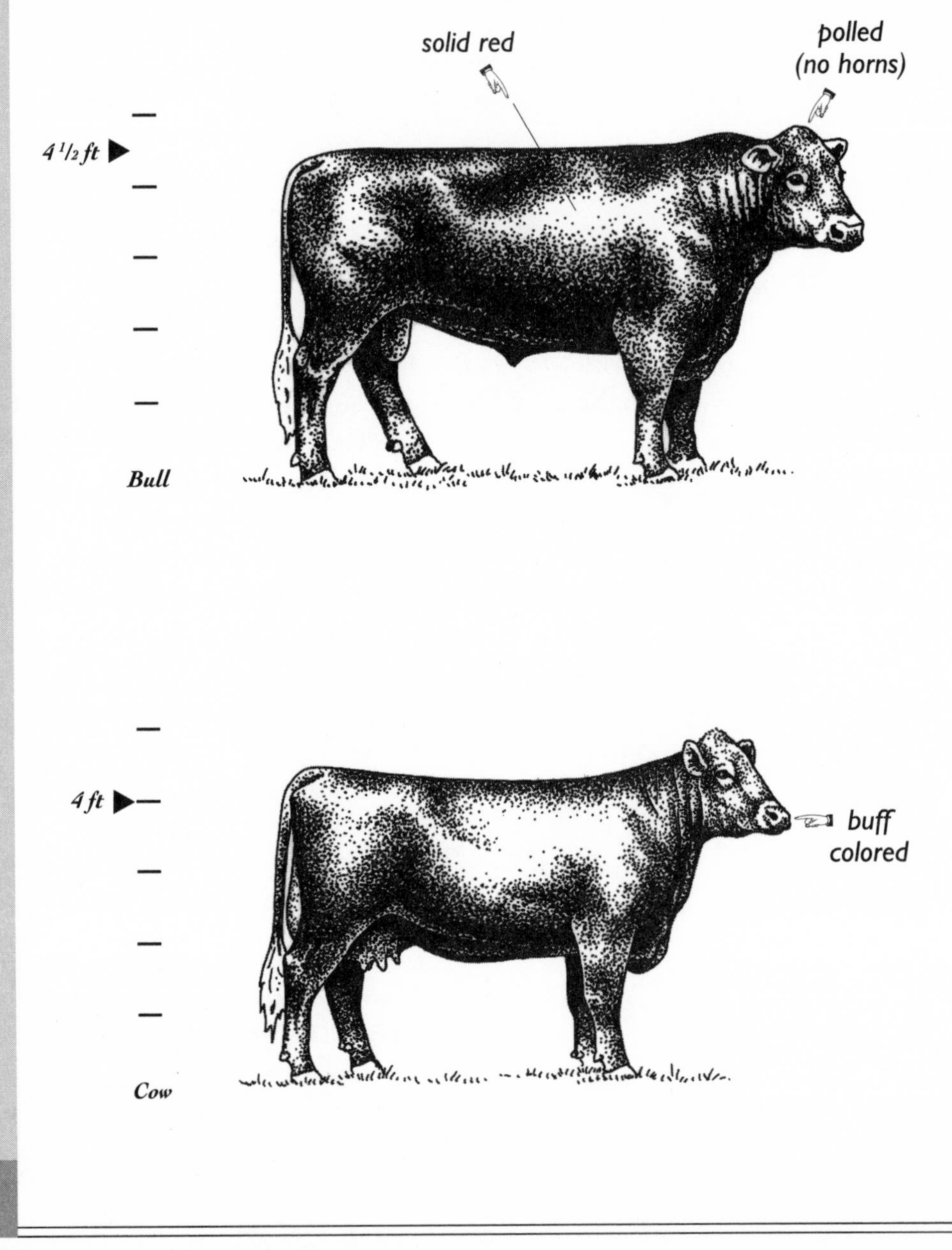

red

Red Poll

- Synonyms: ***Red Polled and Norfolk & Suffolk Red Polled (obsolete)***
- Distribution: ***Sparse***
- Registrations: ***about 1,700***
- Bull's Average Weight: ***2,100 pounds***
- Cow's Average Weight: ***1,450 pounds***

Purpose

Beef and dairy

The Red Poll makes an excellent homestead cow, because of its easy manageability and thrifty habits; it is not persnickety on sparse forage. The Red Poll is also noted for alertness, hardiness, good maternal instincts, lean beef, and a long life (it is among the longest lived of the British breeds). The cows produce milk that is nearly homogenized (that is, it contains small fat globules), which makes it easy to digest. The milk is also ideal for cheese.

Origin

Norfolk and Suffolk Counties, England

The crossing of two ancient breeds—the horned Norfolk and the polled Suffolk—in 1845 gave rise to the Red Poll. The Norfolk was a small, hardy, red, beef cow. The Suffolk was a large, dun, dairy cow with the polled trait genetically dominant. Some researchers believe the Suffolk originated from a dun-colored Galloway line or even a Galloway x Viking cow. Since the first merger, the Red Poll has needed little improvement, already adapting well (as did its ancestors) to eastern England's windy winters and dry summers. The breed was first imported to the United States in 1873, though its parent lines probably made it over with the colonists at some point. The U.S. herdbook formed in 1877.

Cow Fact:
From "Verses from 1929 on" by Ogden Nash:
"The cow is of the Bovine ilk;
One end is moo, the other, milk"
MOO
MILK

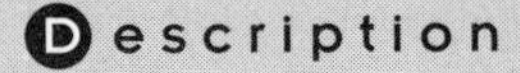

Description

A TALL, CHERRY RED COW.

Usual color is a solid red; small white splotches may appear underneath. Similar to the related Shorthorn, the Lincoln Red differs in its larger stature and consistently solid red body color. Other physical characteristics are buff skin, a long body, tall legs, a straight face, a wide forehead, and short white horns (on those not polled). The Lincoln Red is the largest of the British breeds.

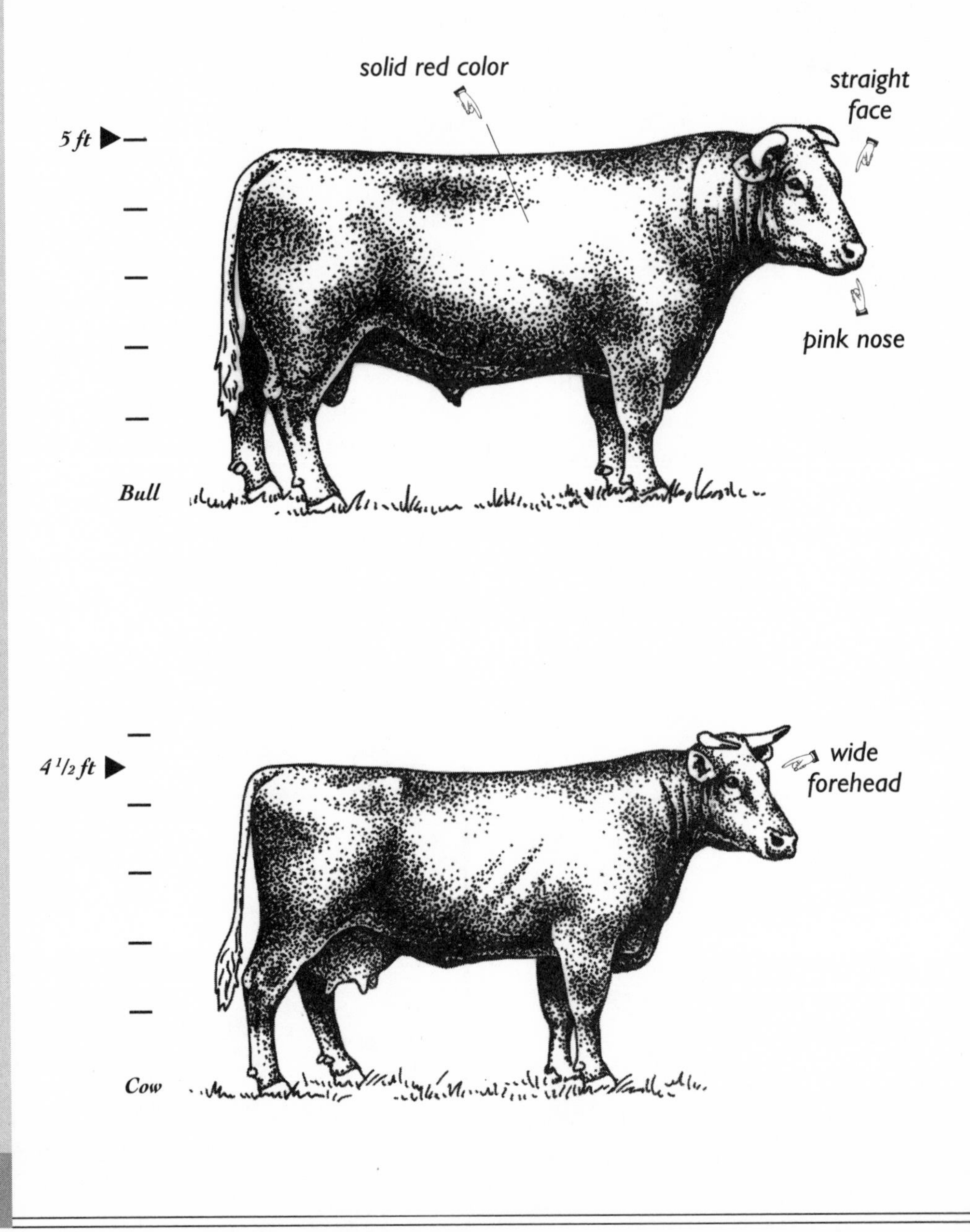

red

Lincoln Red

- Synonyms: ***Lincoln and Red Shorthorn***
- Distribution: ***Limited***
- Registrations: ***N/A***
- Bull's Average Weight: ***1,980 pounds***
- Cow's Average Weight: ***1,540 pounds***

Purpose

Beef and dairy

The Lincoln Red is a dual-purpose breed with emphasis recently placed on the beefiness factor. Said to thrive on modest feeding, the cattle mature rapidly, calve easily, and have a high dressing percent, a docile disposition, and ample milk. The cow is also noted for wet-weather tolerance, which may have developed from grazing along the cold North Sea.

Origin

Lincolnshire, England

The seed for the Lincoln Red was sown between the 8th and the 10th centuries, when Vikings imported cattle to their settlements along the east coast of England. The hardy new cattle intermingled with the indigenous stock and provided the region with a hearty cow for the next few centuries. Only much later, in the 19th century, did the cow develop into its present-day conformation. At that time, farmers started crossing in the massively popular Shorthorn, producing a major influence on the breed's appearance. In fact, the Lincoln Red was known simply as the Red Shorthorn. Still, the two breeds were different enough for the Lincoln Red to break off from the Shorthorn book in 1894 and form its own progressive herdbook. Recently, breeders have crossed in some big European blood (Charolais, Maine-Anjou, Chianina, and Limousin) to increase leanness while retaining the large size.

Cow Fact:

An average cow drinks about 30 gallons of water and eats 95 pounds of feed per day.

Description

A BIG, RED, BLACK-NOSED COW.

The color is a solid deep red; rarely, a few white markings occur underneath. The bulls are a shade darker than the cows, especially around the shoulders. The Danish Red's physical traits include a large angular body; a flat topline; a straight face; short, thick horns; and soft, short hairs, which give the cow a nice, smooth, shiny appearance.

red

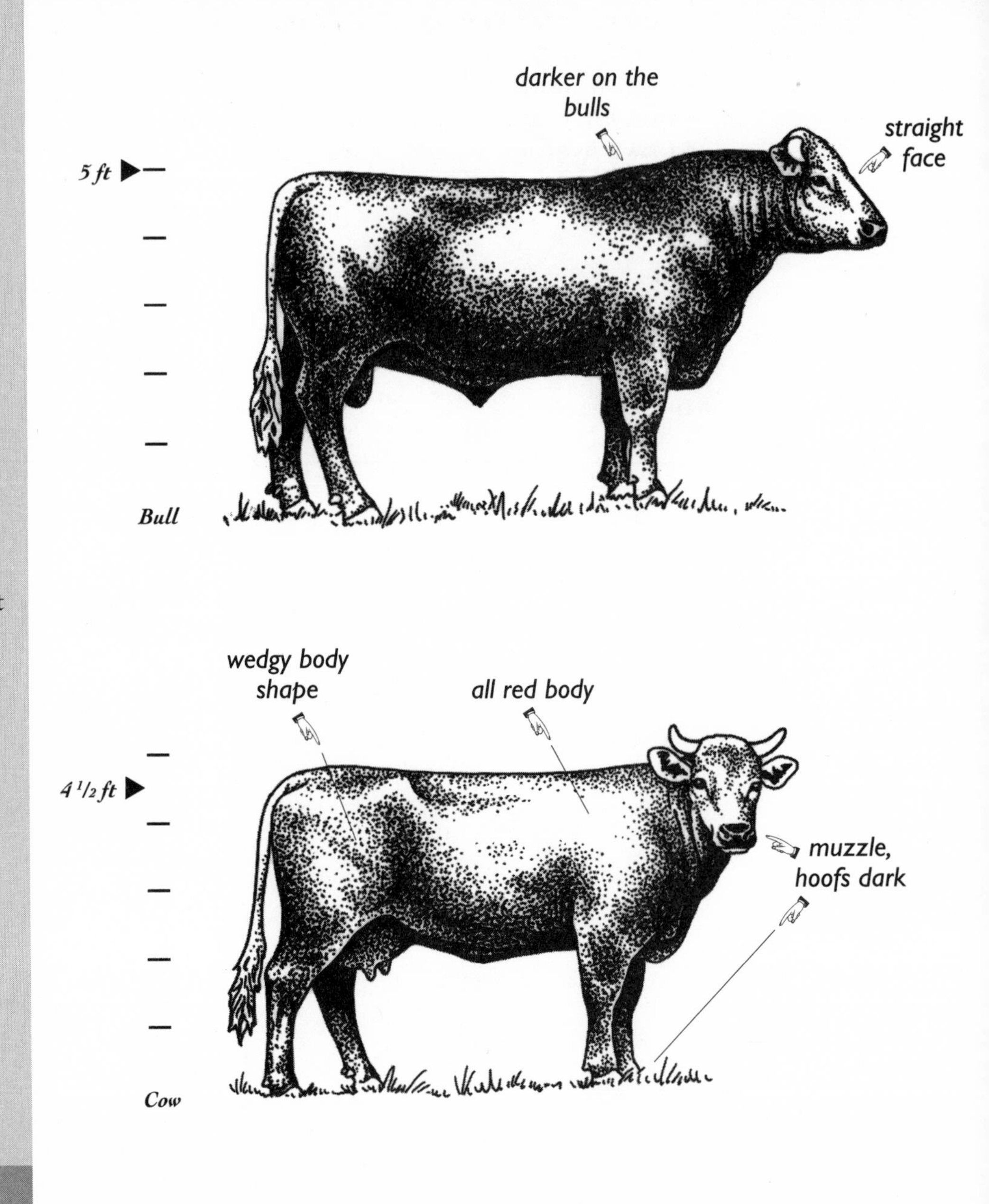

Danish Red

- **Synonyms:** *Red Dane and Rodt Dansk Malkekvorg (RDM)*
- **Distribution:** *Limited*
- **Registrations:** *N/A*
- **Bull's Average Weight:** *2,310 pounds*
- **Cow's Average Weight:** *1,485 pounds*

Purpose

Beef and dairy

In Europe, where it's commonly raised, the Danish Red is a popular and classic dual-purpose breed with the dairy characteristics emphasized. One of the main reasons the Europeans continue to raise the Danish Red instead of some industrial breed is the cow's high-quality milk, which is rich and tasty. Under intensive management, total milk production compares favorably to other breeds. In less intensive applications, the rich milk more than makes up for the smaller volume produced. One Danish cheese available in America is the rich, creamy Havarti. Other traditional Danish Red dairy products are milk, butter, and cow candy.

Origin

Denmark

The Danish Red was developed in the 19th century from the North Slevsig Red, which itself was a cross between the Angln (an old German breed), the local island cattle, and the Bally. The Danish Red was recognized as a breed in 1878 and a herdbook formed in 1885. In 1934, the USDA research station imported 20 head; from there, the breed spread to various parts of the country. An American Red Dane herdbook formed in 1948 but lasted only until 1968. The breed's current situation in the U.S. is unknown.

Cow Fact:

There are an estimated 920 different cow breeds in the world.

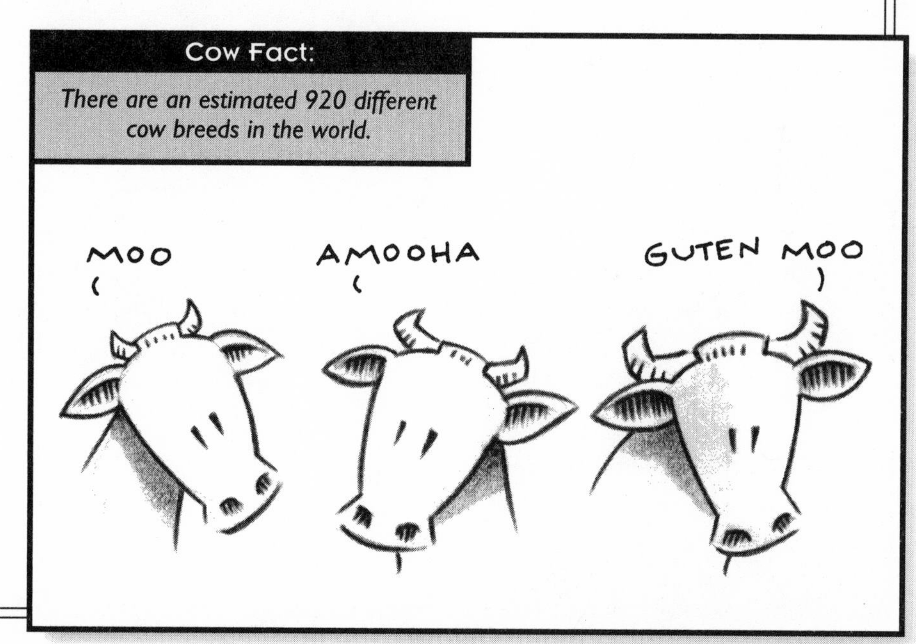

Description

A LONG-HORNED, SHAGGY, WHITE COW.

Black ears, a black nose, and black hoofs contrast with the white coat. Overmarking may appear on an occasional animal (except in the Chillingham herd, where it never occurs); the pattern consists of black spots scattered on the sides, shoulders, or rump and arises from past crossbreeding. Other characteristics include a large foretop; a broad face; a long, narrow body; and a generally handsome appearance. The long horns can point in any direction, though a flat spread is probably the most common.

white with long horns

shaggy coat

long horns

black points

4 ft

Bull

large foretop

3 3/4 ft

Cow

White Park

- Synonyms: ***Ancient White Park, Wild White, White Forest, and White Horned***
- Distribution: ***Critical***
- Registrations: ***about 10 registrations***
- Bull's Average Weight: ***1,100 pounds***
- Cow's Average Weight: ***825 pounds***

Purpose

Beef and wildness

The White Park's main purpose has been to simply exist in a feral state and continue as an ancient historical breed. Even today, the cows run wild in large estate parks in England. The breed is also a valuable genetic source of a divergent line, which means it can provide good heterosis in crosses. Some of the breed's other traits include fast growth on grass pasture, quality lean beef, a long life, adaptability, and a healthy genetic history. The breed is criticized as temperamental, especially the feral cows in the parks.

Origin

Estate parks in Scotland, Wales, and England

From the time of the Roman invasions to around the time of the Norman conquest, the White Park roamed freely in remote British forests. The Normans deprived the animals of their freedom, enclosing them in newly created parks with the justification of safeguarding them. Thereafter, the British nobility hunted the cattle and kept them as their own private quarry. The last noble to hunt cow was King Ed VII in 1877. When the laws changed to allow public hunting, the herds were nearly wiped out. Today, only a few of the original parks still have cows: Cadzow, Chartley, Vaynol, and Chillingham. The Chillingham herd is considered the purest, as no known outside blood has entered the enclosure since the 13th century. The White Park was first shipped to the United States in 1941 to guard against its possible loss during WWII. The American White Park was developed from the White Park, the British White, and the Shorthorn, and has its own breed association.

Description

A SMALL, POLLED, WHITE COW.

A white body and black points or, rarely, red points ("points" refers to ears, eyes, hoofs, and switch). One of the big and obvious differences between the British White and the White Park is that the British White is almost always polled (about 4 percent do possess small horns), while the White Park has a set of long horns. Other physical features include a long, blocky body; a short face; a big foretop; a broad forehead; and big udders. The breed is noted for its good dual-purpose type of conformation.

white

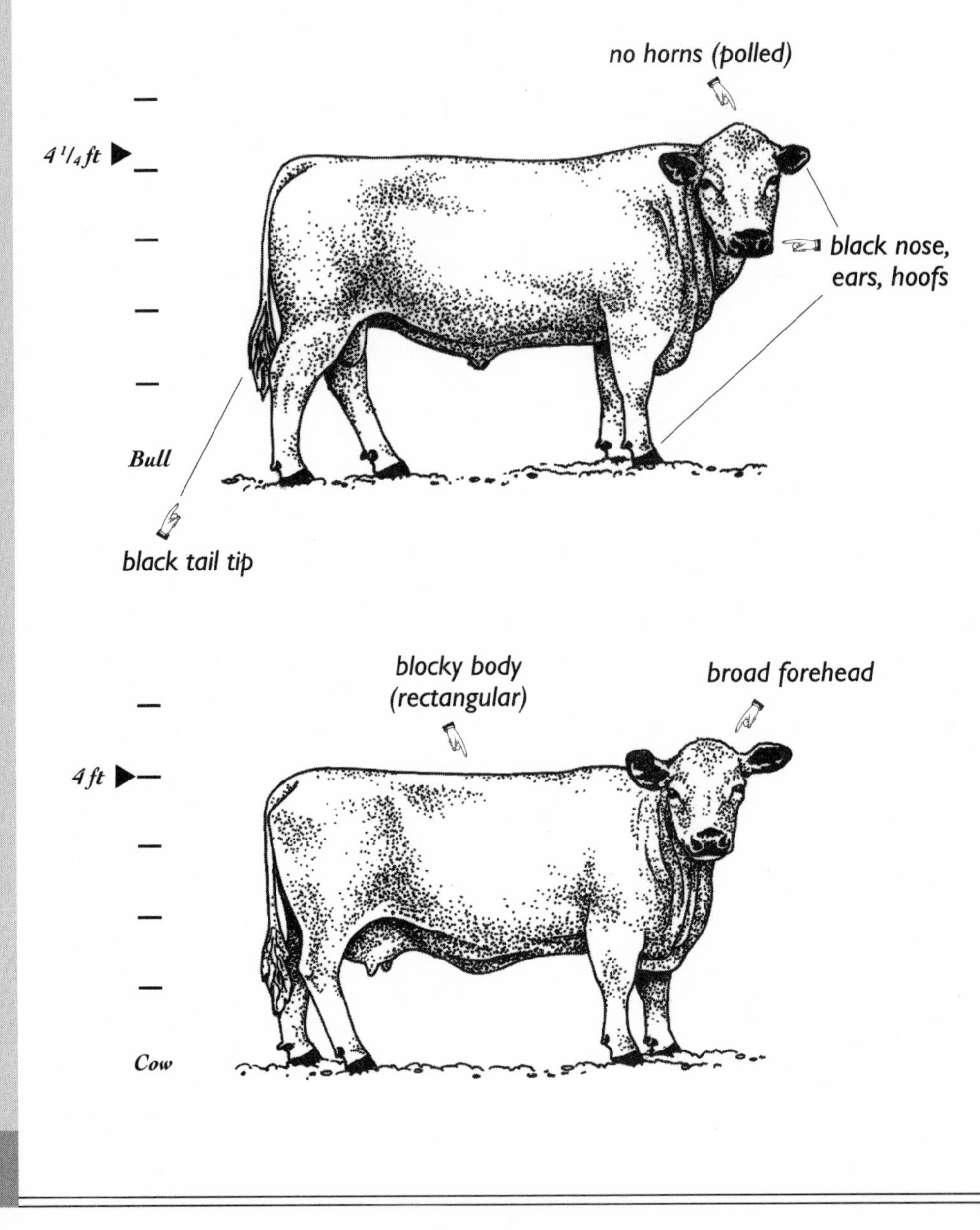

British White

- Synonyms: ***Polled White Park***
- Distribution: ***Sparse***
- Registrations: ***over 2,500***
- Bull's Average Weight: ***1,500 pounds***
- Cow's Average Weight: ***1,000 pounds***

Purpose

Beef and dairy

The British White is a cool-looking cow, ideal for any extensive farming operation, though the breed is also gaining in popularity on all types of American farms for its exceptional attributes. These include hardiness, rangeability, quality lean beef, a gentle disposition, and decent milk yield.

Origin

East Anglia, England

The British White is an old enough breed that conflicting theories exist about its origin. The most plausible is that the breed descended from the Scandinavian mountain cattle; similar cows, called Fjallrus, still ruminate in the remote areas of Sweden. This type of cattle is thought to have garnered a hoofhold in coastal England during the Vikings' plundering days. Another theory has it that the British White descended from the wild White Park by way of farmers who found and emphasized the polled trait. One old record supports this possibility, mentioning a polled herd in the 17th century near Lancaster Abbey, an area known to have formerly had only the wild White Park cows. However, this idea does not hold up to the scrutiny of genetic tests, which show a greater distance between the two breeds. Whatever its source, the British White's numbers decreased enough that breeders crossed in some Galloway, some all-white Shorthorn, and later, some Fjallrus blood. The British White was imported to the United States in 1941, at the same time as the White Park, which caused a bit of confusion in the identity of the two breeds. Later imports arrived in 1976 and 1989. The U.S. British White breed society formed in 1975.

Description

A POLLED, GRAY COW.

The color varies from light gray to dark gray, with the possibility of some white spots underneath. Dark skin pigmentation gives slightly contrasting points on the nose, around the eyes, and on the udder. The Murray-Grey has a stocky, low-set body conformation, similar to that of the original early-1900s Angus. Other features (which are also seen on the Angus) include a smooth, short-haired coat; a short, broad face; and a small, pointy poll.

white

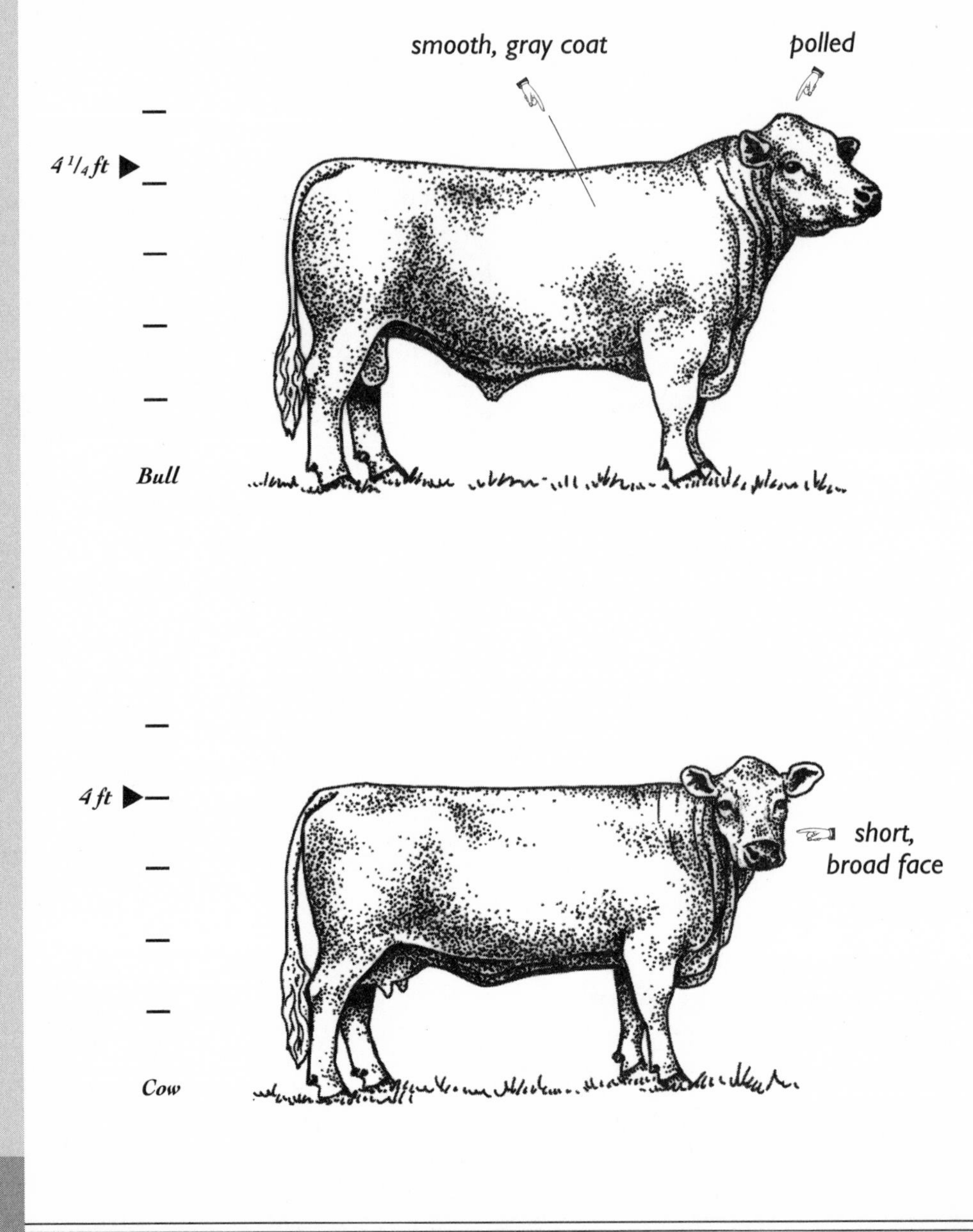

Murray-Grey

- Synonyms: ***Murray and Murray-Gray***
- Distribution: ***Limited***
- Registrations: ***about 1,200***
- Bull's Average Weight: ***2,000 pounds***
- Cow's Average Weight: ***1,200 pounds***

Purpose

Beef

Because the Angus and the Shorthorn formed the basis of the Murray-Grey's bloodline, the breed possesses all the fine British traits, including high-quality beef and early maturity. Diverging from its lineage, the Murray-Grey wears a unique Aussie sun-reflecting gray coat. Other noted traits of the breed include docility, good mothering care, and good waking ability in hot weather.

Origin

New South Wales, Australia

The Murray-Grey breed was born when an Australian rancher's wife expressed a preference for some pretty crossbred gray cattle. An understanding guy, the rancher spared the cattle and thus formed the foundation of the Murray-Grey breed, though he didn't know it at the time. This lucky event happened way back in 1905 up in the Murray River Valley near Victoria, Australia. The gray color of those first cattle came from a cross between an Angus and a light-roan-colored Shorthorn. Thereafter, ranchers placed continual emphasis on the grayness factor, even though they used the black Angus most often in the breeding. The breed only later gained a following; the breed society formed in 1963. The Murray-Grey was imported to the United States in 1969.

Cow Fact:

The differences in coloration between the breeds arose with domestication and the lack of natural selection. In the wild, different-looking and therefore more noticeable animals are the more likely target of predators, while domesticated animals have man's selection and protection, which allow the differences to continue.

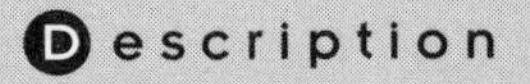

Description

A BIG, WHITE, PINK-NOSED COW.

The color is a creamy white, though the calves are light tan. Overall, the Charolais has an exceptionally balanced beef conformation: the well-developed hindquarters equal the shoulders in breadth. Other features include a short neck; a longer, rougher-looking winter coat than other cows; and a large, gentle-looking face with large eyes. The Charolais is the most common pink-nosed white cow on today's farms.

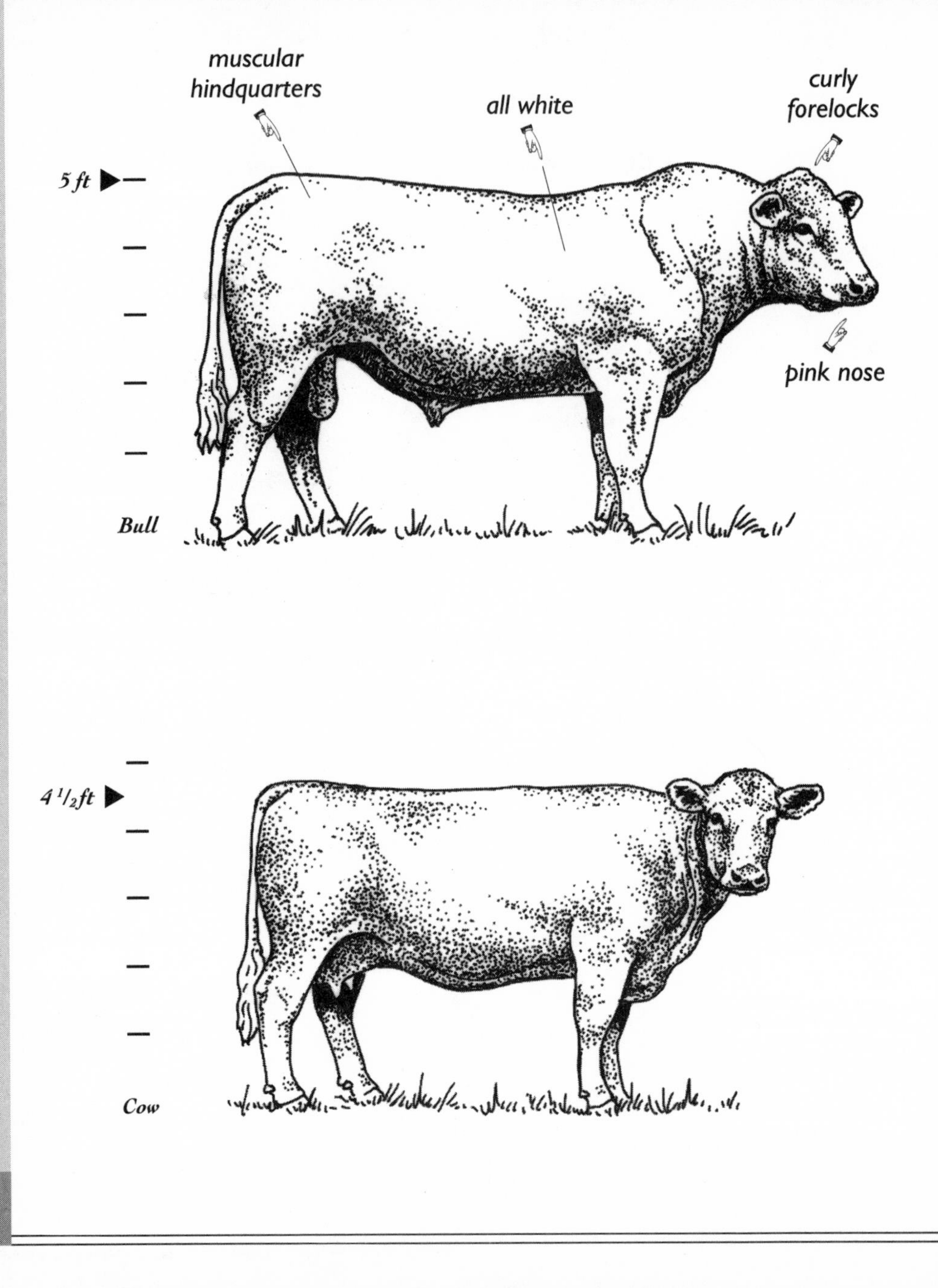

Charolais

- Synonyms: ***None***
- Distribution: ***Common***
- Registrations: ***over 44,000***
- Bull's Average Weight: ***2,585 pounds***
- Cow's Average Weight: ***1,650 pounds***

Purpose

Beef

The hormone-free cow—at least it should be, since the already well-muscled hindquarters don't require any more beefing up. (The hindquarters produce 70 percent of the money cuts: sirloin, tenderloin, filet mignon, etc.) The cow is also noted for its excellent growth thrust, its feeding efficiency, and a wide temperature tolerance. Criticism focuses on the lack of sun-protecting pigment around the eyes, as well as the big head, which can cause calving difficulty.

Origin

Charolles, France

The Charolais comes from central France, near the city of Charolles and its namesake mountain range. One of the oldest and most respected of the French breeds, it was gaining a reputation in the Lyon meat markets for superior quality as early as 1786. From the cow's inception, farmers decided to focus solely on beef traits, rather than on drafting ability or milk yield. Thus they were free to emphasize a balanced body shape (beefing up those hindquarters) rather than big shoulders (necessary in the draft breeds) or refinement (expected of the dairy breeds). Not until the Shorthorn crossing craze struck did some outside blood enter the breed; the new blood did give the animals some earlier maturity. The French herdbook formed in 1864. In the 1920s, the first Charolais arrived in Mexico. For the next thirty years, all Charolais that entered the United States came from Mexico; after that, new regulations on hoof and mouth disease allowed direct importations from Europe.

Cow Fact:

In the ancient Celtic religion the cows of different colors held different meanings: A black cow meant death; a brown cow meant fertility; and white cows symbolized the sun cult.

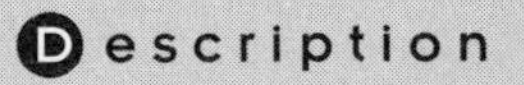

Description

A BIG-RUMPED, WHITE COW.

The color is white or pale gray with all those cow points black (i.e., ears, nose, switch, and hoofs). The Piedmont's claim to fame is its beefy butt. The rump's beefiness comes from an inherited genetic trait called double muscling, which shows up on about 80 percent of the bulls. The rump has the same number of muscles as other cows' rumps, but the Piedmont's muscles are composed of extra-long fibers that give them extra bulk. Other traits include a long body, a smallish head, a dished face, and small, black-tipped horns.

white-big-beefy

big, big beefy butt

dished face

5 ft

dark-colored nose, hoofs

Bull

dark colored tail

Piedmont

- **Synonyms:** ***Piemontese***
- **Distribution:** ***Limited***
- **Registrations:** ***about 1,000***
- **Bull's Average Weight:** ***2,000 pounds***
- **Cow's Average Weight:** ***1,350 pounds***

Purpose

Beef

Formerly a triple-purpose breed, the beefy Piedmont was also used as a strong ox and, surprisingly, a milking cow. Today, the Piedmont is raised solely for beef production and specifically for its double muscling trait. The double muscling increases the total amount of beef by about 10 percent without the use of any nasty hormones, though they say the breed needs intensive management to arrive at its optimum muscle growth. The Piedmont imparts good heterosis in crosses; one popular cross in the Netherlands matches the light-boned Piedmont with the black- and-white Friesian. The cow produces a decent amount of milk, enough to make cheese even after the calf gets its share.

Origin

Upper Po Valley, Italy

The name comes from the Piedmont mountain range in northern Italy. The cow has a mixed heritage: cattle from the plains, the mountains, and other areas all played a part in its mid-19th century beginnings. Some claim the breed has a bit of Zebu blood in it. It seems that the main union, though, was between the golden, long-legged cattle of the valleys and the smaller red-to-straw-colored cattle of the mountains. The first herdbook existed from 1887 to 1891; it started again in 1958.

Cow Fact:

A cow can detect odors up to five miles away.

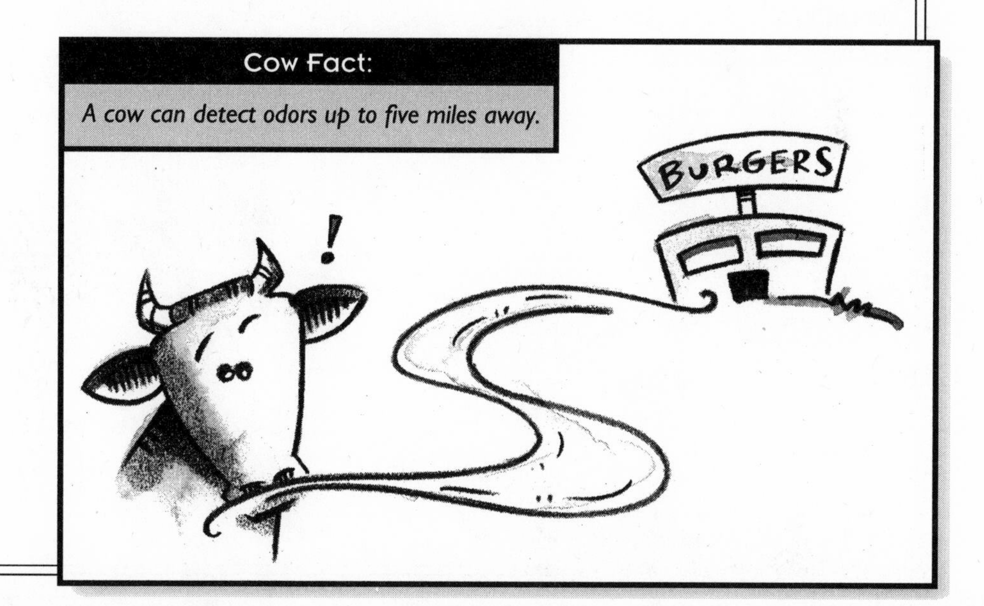

Description

THE BIGGEST COW.

The common coloration is porcelain white with black points, though a steely gray is possible and one specialty line is black. The Chianina is the largest cattle breed in the world. An average bull stands about 6 feet tall at the withers and weighs around 3,000 pounds; the record-weight bull, Donnetto, slammed the scale at a huge 4,300 pounds. The breed presents its massive size well, standing like a chiseled statue. Other features include a fine, narrow head; high-set shoulders; long back; slightly slanting rump; and a generally trim, shapely look.

white and tall

"The tallest cow!"

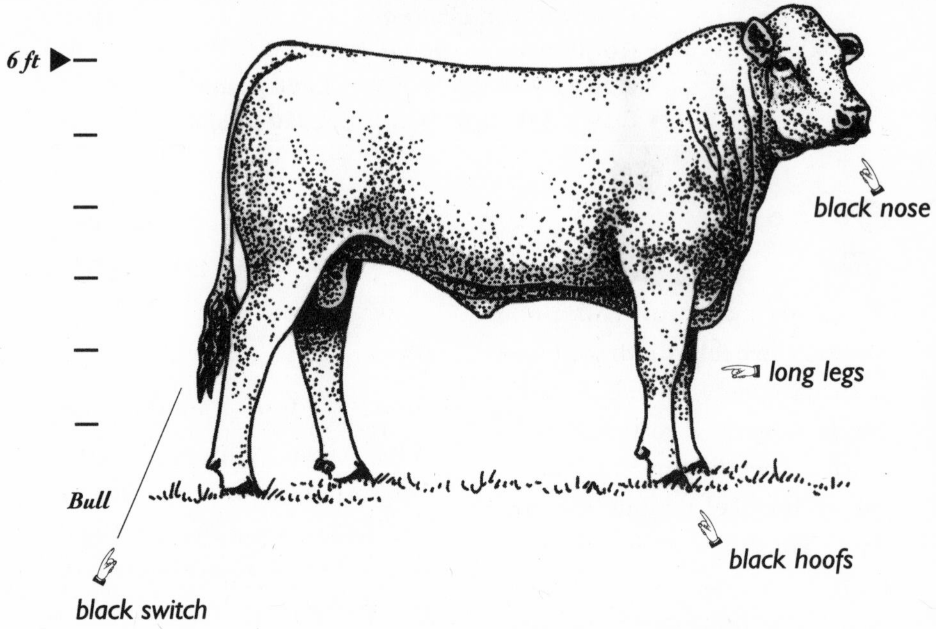

6 ft

black nose

long legs

Bull

black hoofs

black switch

5 ft

Cow

Chianina

- Synonyms: ***Chiana***
- Distribution: ***Common***
- Registrations: ***about 8,000***
- Bull's Average Weight: ***3,000 pounds***
- Cow's Average Weight: ***2,400 pounds***

Purpose

Beef

Formerly a dual-purpose draft and beef breed, today the Chianina focuses solely on beef production and is often used as a terminal sire in crossbreeding programs. The breed is noted for a decent growth rate (4 to 5 pounds a day, though it still takes a while to get up to size), excellent hot-weather adaptation, and few calving problems (its head is small). The cow is a poor milker and has late maturity.

Origin

Chiana Valley, Italy

The Chianina comes from the same region as Chianti wine — the rolling hills of Tuscany. These central Italian hills, it seems, are as good for raising cattle as for growing grapes — the area has nearly ideal pastures 10 months of the year. The cow's present-day conformation came about in the early 19th century, when the big local cattle intermingled with some Podolians. The Italian herdbook formed in 1956. The Chianina was imported to the United States in 1973, during the heyday of the bigger-is-better movement. The cow acted as a model for Michelangelo, who must have thought the Chianina picture-perfect. To practice your Italian, say "key-ahn-na."

Cow Fact:

In 1540 Coronado, coming from Mexico, brought 500 head of cattle onto U.S. soil.

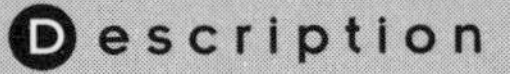

Description

A BIG, ROUGH, WHITE COW.

The color is white or gray with black points; the bull is a shade darker on the shoulders and chest. The cattle have longish, black-tipped horns. Some of the traits that differentiate the Romagnola from the other similar Italian breeds include a small hump; short, stocky legs; a loose hide (especially in the dewlap area); and a coat that grows longer in cold weather. The breed shows a strong, muscular conformation.

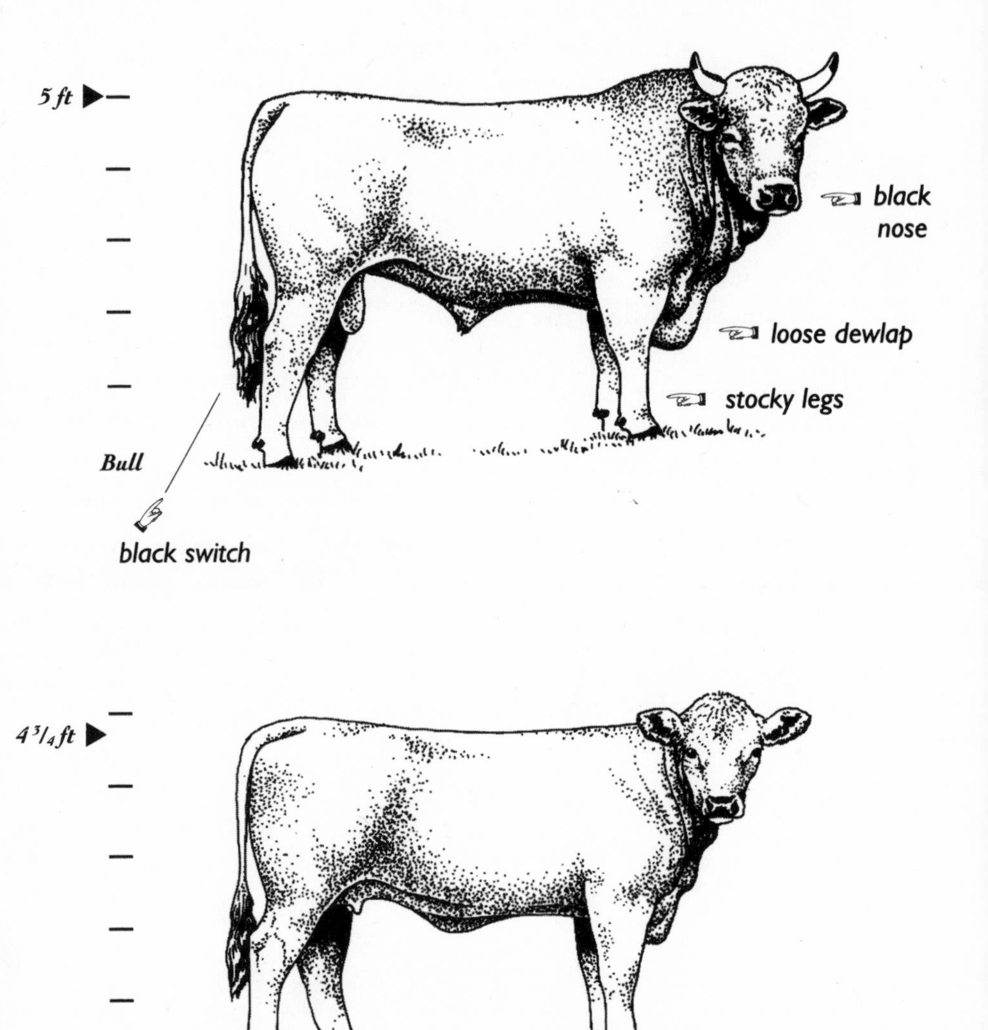

white

Romagnola

- Synonyms: ***None***
- Distribution: ***Limited***
- Registrations: ***about 400***
- Bull's Average Weight: ***2,750 pounds***
- Cow's Average Weight: ***1,600 pounds***

Purpose

Beef

The Romagnola started off as a draft animal, but has been a single-purpose beef cow for the last hundred years. The breed is mainly used as a terminal sire on English beef breeds and other European cattle; it gives good hybrid vigor and provides the genetics for big, beefy calves. Like other Italian breeds, the Romagnola has a good disposition and gives a quality carcass with little waste and a high dressing percentage. The milk yields are decent and the cow adapts to both hot and cold climates.

Origin

Lower Po Valley, Italy

The name comes from the Romagnol Mountains in the lower Po Valley of northern Italy (the Piedmont claims the upper Po). The rough terrain of these mountains in the Apennine Range hides the fact that the land is rich and fertile.

The region's farmers are believed to have kept the breed pure throughout the years, mixing no outside blood with the original stock; the Romagnola is thought to be the purest Italian Podolian. In fact, genetic mapping verifies this, showing the breed to be genetically distant from its neighbors. Between 1850 and 1880 there was some mixing with the Chianina. The United States breed society formed in 1974.

Cow Fact:

There are approximately 350 squirts in a gallon of milk.

A BIG WHITE COW.

The color is white or grayish white; the bulls are usually a shade darker than the cows. Like the other Italian breeds, the Marchigiana has small, black-tipped horns and black points on the nose, ears, and hoofs. The breed possesses a strong beef conformation: a straight topline; a long, level rump; and a deep, cylindrical body. The Marchigiana differs from other Italian breeds in its slightly darker body color and its moderate-length legs. Its cut—or appearance—is right between the refined, longer-legged Chianina and the longer-horned, rougher-looking Romagnola.

white

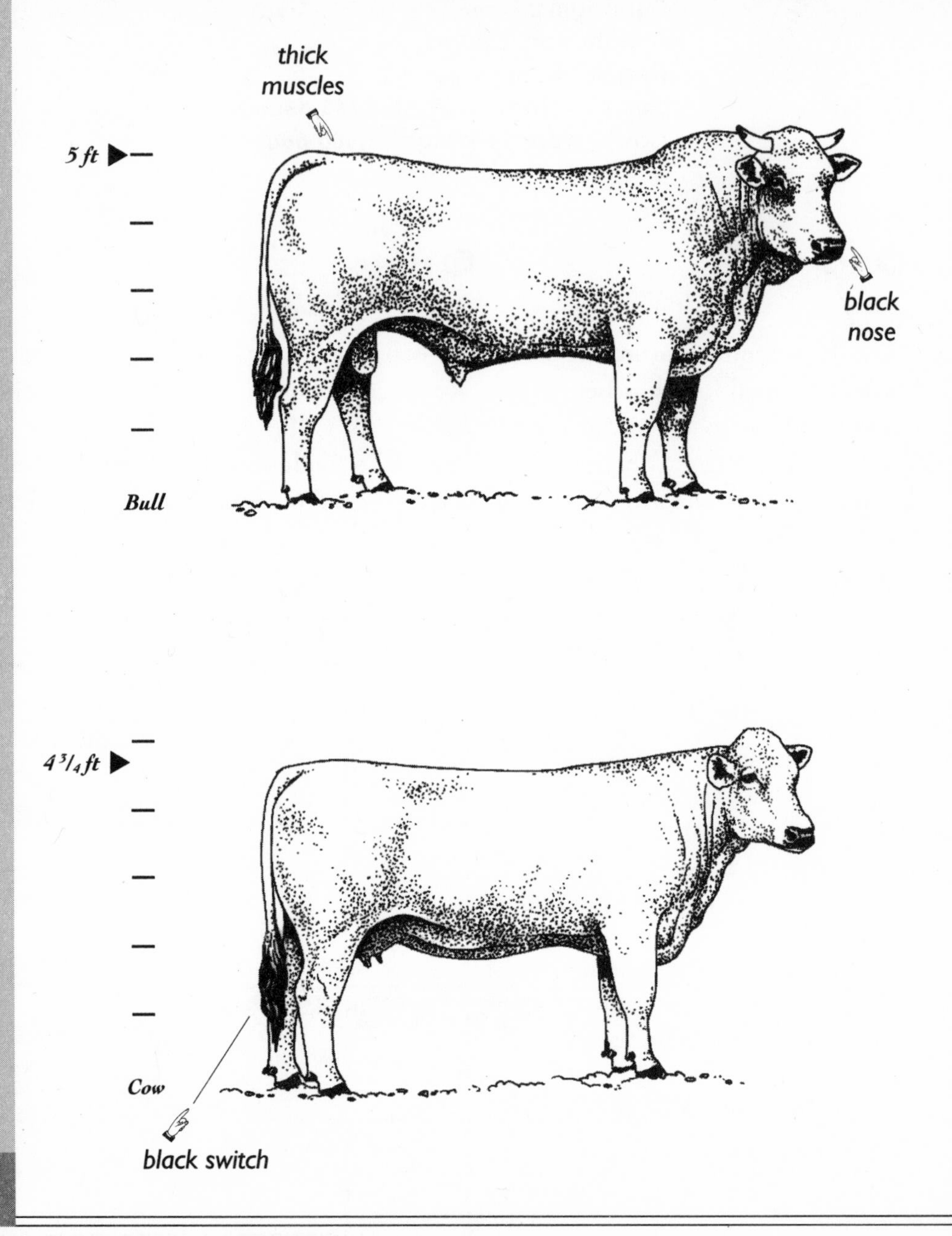

Marchigiana

- Synonyms: ***Marky and Del Cubante***
- Distribution: ***Limited***
- Registrations: ***about 75***
- Bull's Average Weight: ***2,400 pounds***
- Cow's Average Weight: ***1,300 pounds***

Purpose

Beef

In the past, the Marchigiana was mainly a draft animal; today it is raised strictly for beef. The Marchigiana is known for heat adaptation (the white hair and black skin help it in the sun), insect resistance, and a hardy nature in a variety of climates and terrains. Other traits include strong bones, a good growth thrust, decent milking ability (around 7% butterfat content), easy calving, and a gentle disposition.

Origin

Central Italy

The Marchigiana comes from the central Italian region bordering the Adriatic Sea. The breed had its beginnings in the 5th century: at that time, northern barbarians brought Podolian cattle into the region and allowed them to mix with the indigenous stock. Not until the mid-19th century, when Italian farmers started crossbreeding the cows, did the present-day conformation take shape. The first crosses were made with the Chianina; later, in the 20th century, farmers used the Romagnola. All crossbreeding stopped in 1932; the breed has been kept pure since then. The Italians organized a herdbook in 1957. Today, the Marchigiana constitutes 45 percent of Italy's cattle population. First imports to the United States arrived by way of Canada in 1973. The name is pronounced "mar-key-jahna."

Cow Fact:
A seven-year-old cow.
1 2 3 4 5 6 7

Description

A DROOPY-EARED, HUMPED COW.

The body coloration is variable, though either a solid gray or a solid red shows up most frequently. Other possible colors include black, brown, white, and spotted. Whatever the color, it fades to a lighter shade underneath. The identifying hump usually stands straight up. The hide is loose and saggy, especially in the dewlap, the sheath, and the big, droopy ears. Other features include a long, concave face; a sloping rump; an easy gait; and straight, medium-length horns. One more trait is oral: the cows grunt rather than moo.

baggy & humped

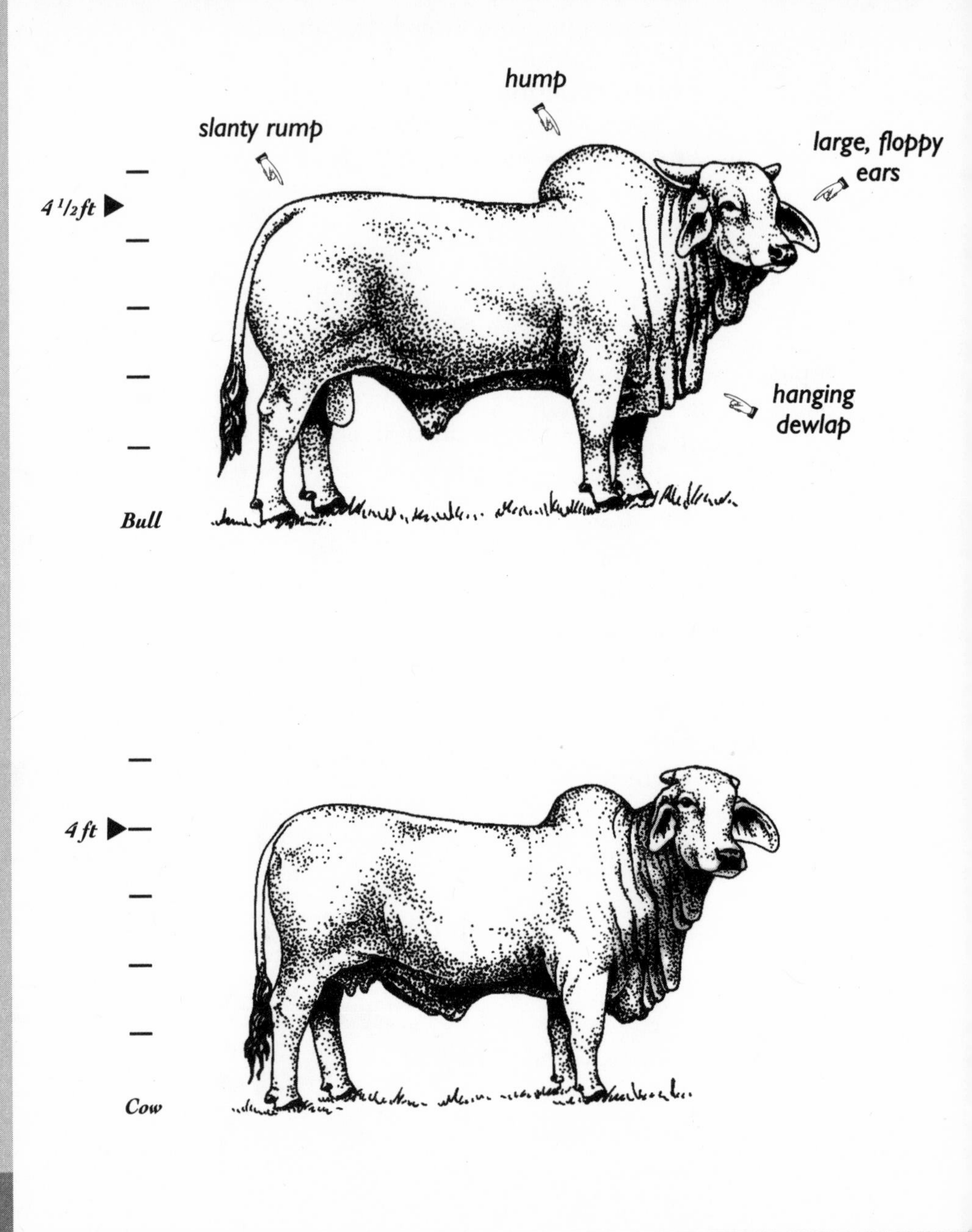

Brahman

- Synonyms: ***American Brahman and Brahmer (the south)***
- Distribution: ***Common***
- Registrations: ***over 16,000***
- Bull's Average Weight: ***2,090 pounds***
- Cow's Average Weight: ***1,320 pounds***

Purpose

Beef

Brahman's *raison d'etre* is excellent heat and insect resistance. This adaptation makes the breed ideal for the hot, humid southern states and explains its popularity in crosses. The Brahman's short hair, loose skin, and ability to secrete insect-repelling sebum from its skin help it cope with heat and insects. Other traits include good rustling ability, quick growth on pasture, long life, and rich milk. The Brahman is said to appreciate good handling and is criticized for poor beef quality.

Origin

U.S.A., developed from Indian breeds

Although India has over 30 native indicus (humped cattle) breeds, the American Brahman started from only four valley breeds: the Gir, the Krankrej, the Krishna Valley, and the Ongole (also called the Nellore). Normally, the cows were sacred and could not be exported for consumption. The British government, ignoring local customs, presented a few of these cattle as a gift to the United States government in 1854. During the next 50 years, the English allowed more cattle to be exported. The original gift proved to be fortuitous for the U.S.: the new hot-blooded cows made beef ranches in the hot, humid South profitable for the first time. The Brahman became the first developed American cattle breed and a breed society formed in 1924. Today, almost all the cattle in hotter climates around the world depend on at least some indicus blood.

Description

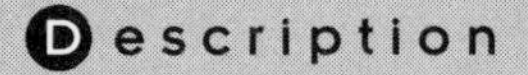

A BAGGY, BLACK COW.

The hair and skin color are an Angus black and the body has a Brahman's sagginess. The Brangus is always naturally polled. Other physical features include a short, dished face; a sloping rump; legginess; and big, droopy ears. The Brahman's big hump does not transfer in the cross.

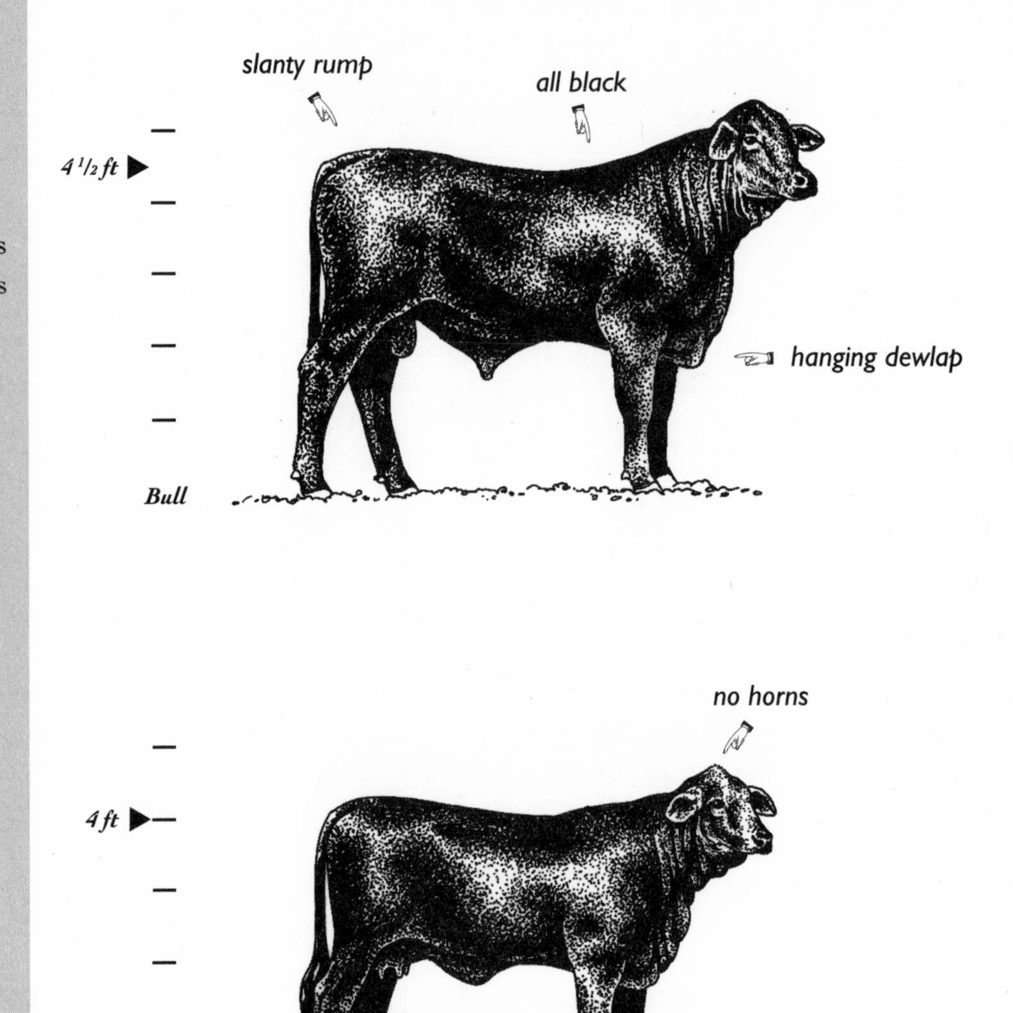

Zebu cross

Brangus

- Synonyms: ***None***
- Distribution: ***Common***
- Registrations: ***about 32,000***
- Bull's Average Weight: ***2,145 pounds***
- Cow's Average Weight: ***1,450 pounds***

Purpose

Beef

The Brangus is a Brahman cross that unites heat tolerance and quality beef in one breed. The cross also has the Angus's cold tolerance, which makes it more cold-weather adaptable than other Brahman crosses. Brangus traits include good mothering ability, vigorous and hardy calves, a generally poor temperament, and good adaptability to the large beef ranches of the southwestern U.S., where it is often found.

Origin

Louisiana, U.S.A.

In 1932, the United States Department of Agriculture tried the first Angus and Brahman crosses at their station in Jenerette, Louisiana. Not until later, in 1942, did ranchers in Oklahoma and Texas see to most of the breed's development. The standard crossing procedure is a complicated mess, and works out something like this: take a purebred Brahman, cross it with a halfblood Brahman x Angus animal, which makes a 3/4 Brahman–1/4 Angus cow. Cross this cross with a full-blooded Angus for the final pure Brangus cow of 3/8 Brahman x 5/8 Angus. The U.S. Breed Society formed in 1949.

Cow Fact:

The lighter under-color seen on some cows is said to act as a natural camouflage by highlighting the shadow underneath the cow, which thereby renders the cow more two-dimensional and harder for a predator to see.

hmmm-
which way
did he go?

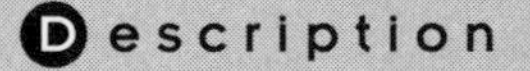

Description

A BAGGY, BESPECTACLED, WHITE-FACED COW.

The most common coloring is a Hereford-like white face, a red body, and red encircling the eyes like a mask. Other color variations include solid red, solid black, and skewbald. The Brahman heritage is seen in the Braford's baggy dewlap, hanging sheath, legginess, light build, dished face, and short, stocky horns.

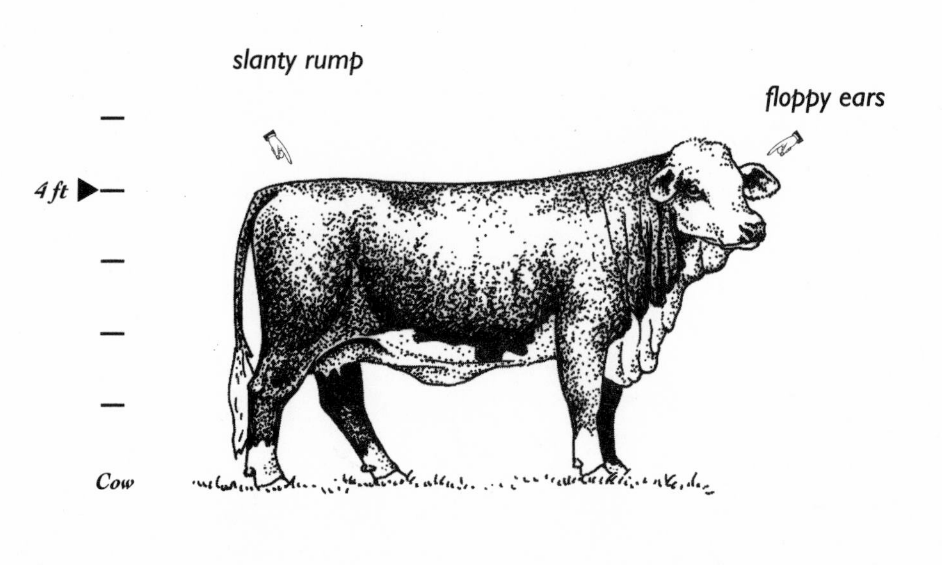

Braford

- Synonyms: ***Herebu (Argentina)***
- Distribution: ***Sparse***
- Registrations: ***about 2,200***
- Bull's Average Weight: ***1,750 pounds***
- Cow's Average Weight: ***1,250 pounds***

Purpose

Beef and rodeo

The Braford is a beef cow for the hot and humid climates. The Hereford parentage gives decent beef qualities and cutability. The Brahman blood gives insect and heat resistance as well as some orneriness. This breed is often seen in rodeo bull-riding contests; its orneriness serves it well when it's bucking its riders.

Origin

Florida, U.S.A.

The Braford is one of those recent breed crosses that ranchers can make themselves. This means ranchers have the option of varying the proportion of Hereford to Brahman blood, though they usually keep the ratio of Brahman at around $^{3}/_{8}$ to $^{1}/_{2}$ in the final cross. The standard Braford started on the Adams Ranch in Fort Pierce, Florida. After raising European cattle, the ranch decided to change to a Brahman cross, hoping it might work better in the heat and still retain the good British beef quality. Right off the bat, the ranch gave its new breed a broad genetic base, using 400 Hereford bulls on Brahman cows and forming 5 separate family lines. They set up some basic rules: test yearling weight, give no assistance during calving, do no creep feeding and no color selection (except, when all else is equal, they prefer eye patches), and basically don't interfere in natural selection. All this gave a cow that requires a minimum of management and is durable on the large ranches. The ranch continues to raise the Braford and currently has over 10,000 mother cows. Another Braford line was developed separately in Australia.

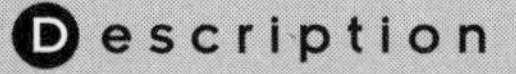

Description

A BIG, DROOPY, RED COW.

The most common color is red, with or without white splotches. A few other color varieties include red-brown, dun, and roan. The Brahman influence is visible in the Beefmaster's saggy dewlap, big sheath, long ears, dished face, and sloping rump. Overall, the Beefmaster has a good beef conformation, especially noted for the well-developed hindquarters.

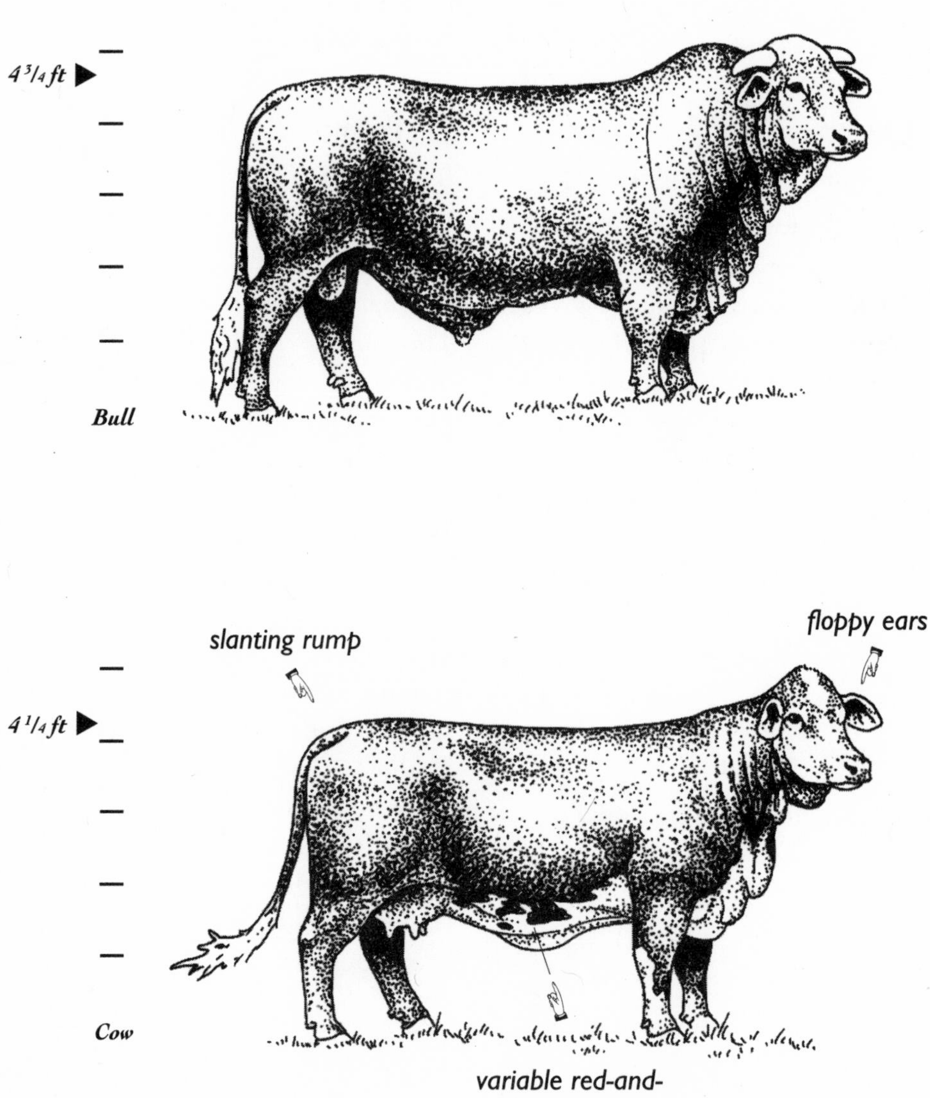

Zebu cross

Beefmaster

- **Synonyms:** ***None***
- **Distribution:** ***Common***
- **Registrations:** ***Over 36,000***
- **Bull's Average Weight:** ***2,640 pounds***
- **Cow's Average Weight:** ***1,650 pounds***

Purpose

Beef

First developed and now raised for one thing only: beef production. The Beefmaster is primarily found in the region for which it was developed and for which it is so well adapted, on the large, hot, dry southwestern ranches of Texas and Oklahoma. Its popularity comes from its combination of the Brahman's heat and insect resistance with the English breeds' superior beef quality.

Origin

Lasater Ranch, Texas, U.S.A.

Texan Tom Lasater developed the Beefmaster from a cross of 25 percent Hereford, 25 percent Shorthorn, and 50 percent Brahman. His ranch began raising Brahmans in 1908 and kept at it 'til the 1930s, at which time he decided he needed more beef in his cows. Approaching the task in a completely analytical fashion, he named the following six traits as the most important for his new beef cow: 1) disposition, 2) fertility, 3) weight, 4) conformation, 5) hardiness, and 6) milk production—this ensures the calves get enough food for quick growth. No emphasis was placed on inconsequential traits like coat color: thus the wide variety in coloration mentioned above. Plus, Lasater let all the action take place on the range, and he did not use only one main sire, so the new breed ended up with a fairly large genetic base.

Cow Fact:

Texas Longhorns never dominated the northern and central plains because cold limited their range to the warmer south.

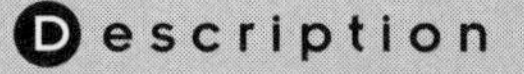

Description

A SAGGY, SOLID CHERRY RED COW.

The cow is almost always solid red, though white markings may show up underneath, on the udder, and on the switch. Physical traits include red-pigmented skin; short, slick hair; large, floppy ears; a wide forehead; a slanting rump; and short horns. The Santa Gertrudis stands fairly tall on strong legs, and its body is cylindrical.

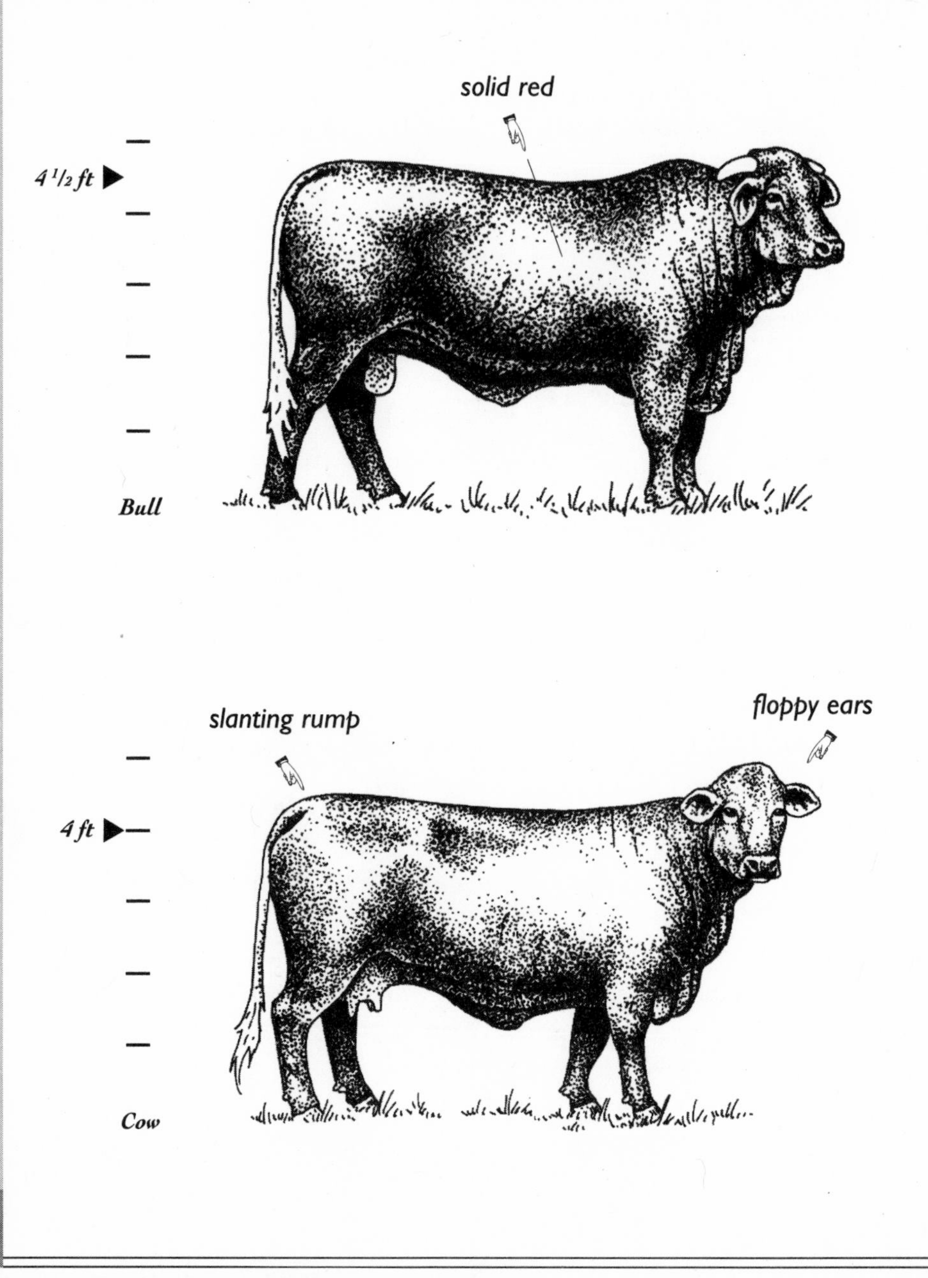

Zebu cross

Santa Gertrudis

- Synonyms: ***Gertie***
- Distribution: ***Common***
- Registrations: ***about 6,000***
- Bull's Average Weight: ***2,145 pounds***
- Cow's Average Weight: ***1,430 pounds***

Purpose

Beef

The Santa Gertrudis is a single-purpose beef breed developed for the arid climate of the southwestern United States. Like the other Brahman crosses, the Santa Gertrudis has decent heat and insect resistance, good cutability, a good growth thrust, and decent beef quality. The breed is criticized for poor fertility.

Origin

King Ranch, Texas, U.S.A.

Some like to think of Santa Gertrudis as the patron saint of cows. The name actually comes from a Spanish land grant in Texas, where Captain Richard King started his huge King Ranch. He also developed the Santa Gertrudis breed, which he named in honor of the gift from Spain. Captain King had his pick of the nation's breeds, including the British beef breeds, the Texas Longhorn, the Brahman, and even the Africander. He decided the ranch needed a new cattle breed specifically suited for the southern U.S.: one that could stand up to the Texas heat and still make a decent steak. So the rancher used his many breeds to start a cross-breeding program. From all the possible combinations, King finally decided on a bull, from a Shorthorn and Brahman cross, named Monkey. This bull was crossed with the top cows of the ranch's huge supply to make the Santa Gertrudis's foundation stock. After much further breeding, the Santa Gertrudis ended up with 5/8 Shorthorn and 3/8 Brahman blood. Today, over 60,000 head of Santa Gertrudis still graze on the 825,000-acre ranch. The breed society and herdbook formed in 1951.

Cow Fact:

Old cows in India have their own nursing homes.

Description

A RECTANGULAR-FACED, RED COW.

The color varies from a light cherry red to a dark mahogany, with the possibility of white marking underneath. From its Africander blood, the Barzona gets many of its physical features, including a rectangular face; down-curving, small, oval horns; a sloping rump; and a slightly raised hump, which shows its Brahman blood. Additionally, the cow has skinny legs that are actually quite strong; they just have less excess tissue surrounding the tendons.

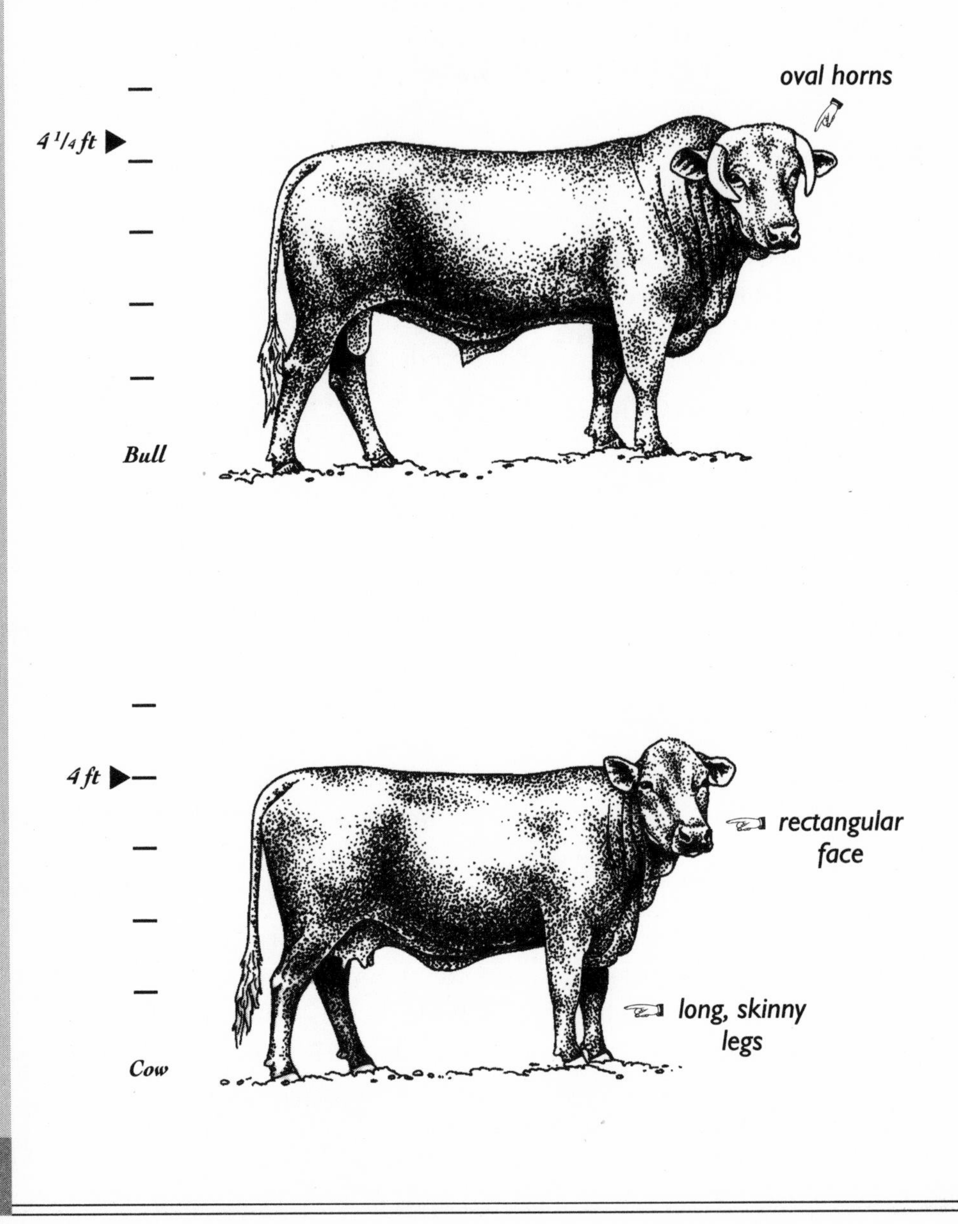

Zebu & Sanga cross

Barzona

- Synonyms: ***None***
- Distribution: ***Sparse***
- Registrations: ***about 1,300***
- Bull's Average Weight: ***1,650 pounds***
- Cow's Average Weight: ***1,240 pounds***

Purpose

Beef

The Barzona was developed solely for the intermountain region of the southwestern United States and Mexico, where the breed specializes in foraging on any available grasses or plants, even the prickly pear cactus. In order to find this forage and to keep on chewing its cud, the cow makes use of its excellent walking ability (on those skinny legs), as well as its decent resistance to heat and insects. The cows are also expected to be exceptional mothers, which means they have to keep their calves out of trouble while grazing in rugged terrain.

Origin

Bard Ranch, Arizona, U.S.A.

F. N. Bard developed the Barzona between 1945 and 1968 on the Bard Ranch near Kirkland, Arizona. The main breeder was a hired gun, the geneticist Elliot Humprey, who set about creating a breed for this region in a thoroughly scientific manner. Humprey decided the ideal cow attributes would include wild grazing, heat/insect tolerance, strong feet and legs, mothering ability, and carcass quality. He chose British breeds (Angus and Hereford) for beef quality, the Santa Gertrudis for heat and insect tolerance, and the Africander for rustling ability and hardiness. All breeding took place under natural conditions on the open range; Humprey kept detailed performance records of all the cattle, from which he determined the best animals for breeding. The Barzona ended up with this convoluted breeding mixture:

[Angus x (Africander x Hereford)] x [Santa Gertrudis x (Africander x Hereford)].

Quite simply, this works out at about 25 percent of each breed. A breed society formed in 1968. The name Barzona is a combination of Bard and Arizona.

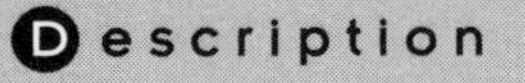

Description

A FUNNY-LOOKING BROWN COW OR A BIG-HUMPED COW.

Frequently dark brown, though lighter shades are possible. Long, shaggy hair covers the head and hump. The hump itself consists of massive muscles supported by the backbone, which reaches up to provide the foundation. European Bison, Woods Bison, and female Plains Bison all have smaller humps than the male Plains Bison. Other identifying traits include short, incurving horns; a long goatee; and a narrow waist. Also, Bison face into driving wind, rain, or snow, unlike *Bos* (domestic) cows, who turn their backs on such nasty stuff.

big hump
6 ft
narrow waist
large head
Bull

5 ft
Cow

Bison

- Synonyms: ***Buffalo, American Buffalo, and Tatanka (Lakota)***
- Distribution: ***about 80,000 animals are found in United States national parks and on ranches.***
- Bull's Average Weight: ***1,600 pounds***
- Cow's Average Weight: ***1,000 pounds***

Purpose

Beef, hunting, and scenic aesthetics

The Bison's original purpose was just being part of the natural ecosystem. Today, over 90 percent are raised on ranches for beef or amusement. When raised for beef production, the Bison has decent traits, including superior rustling ability (they can dig under snow to find forage), longevity (2 to 3 times that of the *Bos* breeds), excellent maternal ability, and top-quality lean beef that makes great-tasting hamburgers.

Origin

North America

At the time of the last Ice Age, the Bison was prevalent throughout much of Asia and Europe. It was these animals, about 25,000 years ago, that first crossed over the Bering Sea land bridge. With this short jump, the Bison became the sole member of the taxonomic family Bovidae in North America; that is, until the European milking cow's arrival. Migrating southward from Alaska and the far north, the Bison became established in the woods of Canada and on the plains of the American West. The herds expanded to an estimated 30 million head, which is enough for the cliche "as far as the eye could see" to be true. The Bison's decline started in the early 19th century with an eradication program which took only 50 years to nearly accomplish its goal. Fortunately, a few thousand head survived the slaughter; most of these ended up in national parks. The plains were filled by feral Longhorn cattle and European settlers bringing with them their own English cows. Ironically, the Bison is once again finding a place on ranches right dab in the middle of its former range.

Beefalo

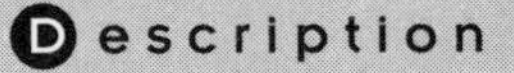

AN INDEFINITE, FUZZY COW.

A shaggy, dense coat of hair, twice the density of that on a *Bos* cow, is about the only consistent feature of the Beefalo. The other characteristics are variable since any breed of cow might be crossed with the Bison, resulting in a full range of colors and conformations. When breeders do the cross, they usually keep the Bison blood to a minimum, meaning the hump and other Bison traits stay hidden.

Beefalo

- Synonyms: ***American Breed and Cattalo***
- Distribution: ***Sparse***
- Registrations: ***no registrations***
- Bull's Average Weight: ***1,500 pounds***
- Cow's Average Weight: ***1,000 pounds***

Purpose

Beef

Breeders are looking for a combination of the Bison's hardiness and the *Bos's* growth thrust and docility in one animal. Ideally, the Beefalo will be a weather-tolerant cow, a natural rustler, a lean-beef producer, and a gentle enough animal to work with on a ranch.

Origin

U.S.A.

Charles Goodnight, a Texas Ranger, first bred the Cattalo, though his experiment met with only limited success. In the years since, other ranchers have made attempts at crossing the Bison against various *Bos* breeds, but the first generation always seemed to have a problem with infertility. Breeding in a higher percentage of *Bos* blood mitigates the problem somewhat. For example, the American Breed (1/2 Zebu and 1/8 each of Charolais, Hereford, Shorthorn, and Bison) works a bit better with the lowered Bison percentage; but the traits of the Bison get lost in the mix. Because of the fertility problem, the cross is not too popular with ranchers. At best, the Beefalo might end up as a novelty breed or a terminal cross.

Cow Fact:

People first domesticated cows about 5,000 years ago.

get up and do something you lazy cow

About 5,001 years ago:

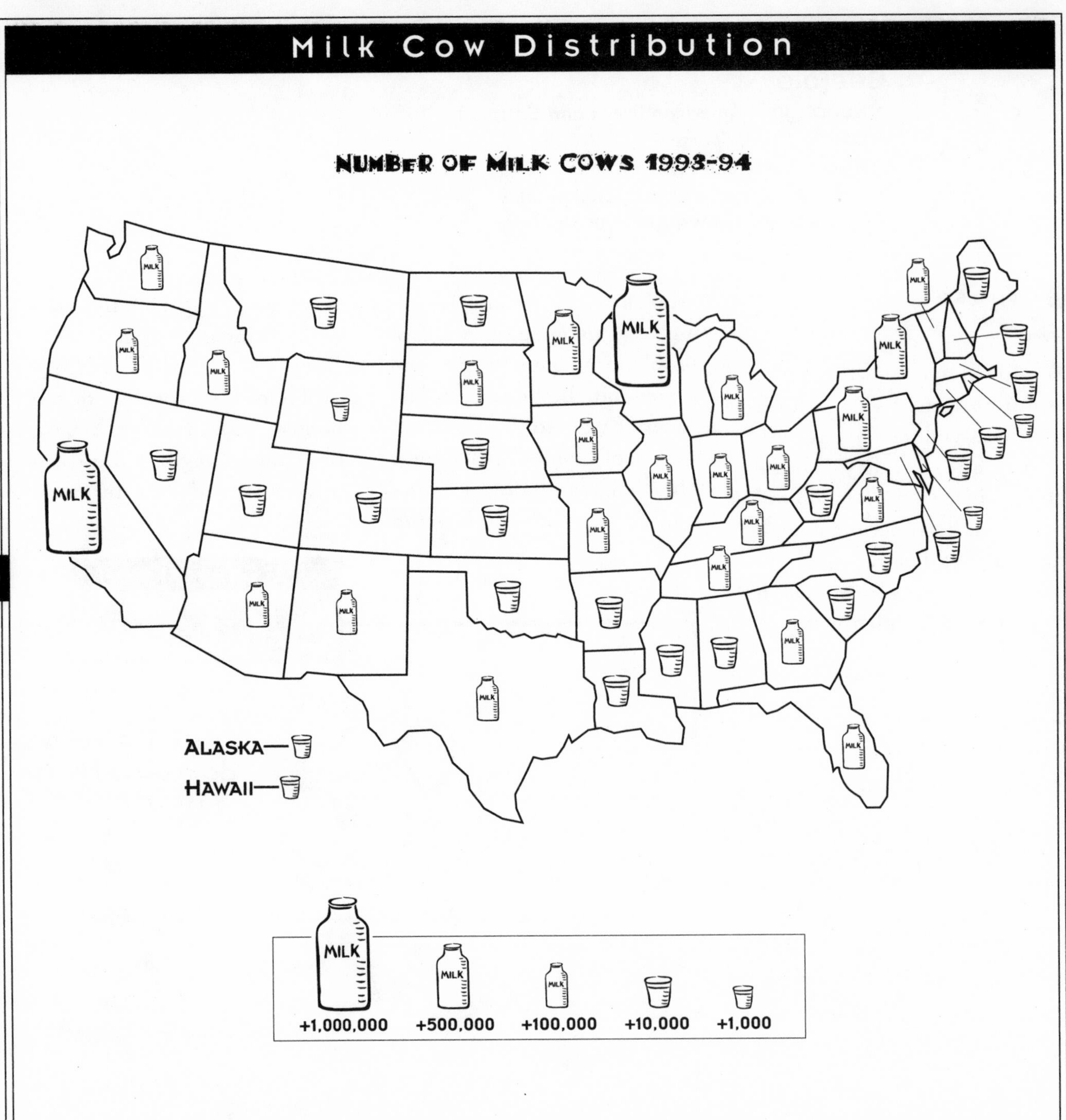
NUMBER OF MILK COWS 1993-94
MILK
ALASKA—
HAWAII—
+1,000,000
+500,000
+100,000
+10,000
+1,000

Ranking of Dairy States

The ranking of states in terms of number of milk cows. The ranking of total milk production of the states is different, with California coming in first.

NATIONAL AGRICULTURAL STATISTICS SERVICE 1993-94

RANK	STATE	MILK COWS	RANK	STATE	MILK COWS
1	WISCONSIN	1,494,000	26	MARYLAND	92,000
2	CALIFORNIA	1,235,000	27	NORTH CAROLINA	90,000
3	NEW YORK	718,000	28	UTAH	86,000
4	PENNSYLVANIA	639,000	29	COLORADO	81,000
5	MINNESOTA	609,000	30	LOUISIANA	79,000
6	TEXAS	402,000	31	KANSAS	78,000
7	MICHIGAN	328,000	32	NEBRASKA	77,000
8	OHIO	294,000	33	NORTH DAKOTA	68,000
9	IOWA	265,000	34	ARKANSAS	61,000
10	WASHINGTON	261,000	35	MISSISSIPPI	57,000
11	IDAHO	208,000	36	MAINE	40,000
12	MISSOURI	197,000	37	ALABAMA	37,000
13	FLORIDA	176,000	38	CONNECTICUT	33,000
14	KENTUCKY	168,000	39	MASSACHUSETTS	29,000
15	NEW MEXICO	165,000	40	SOUTH CAROLINA	28,000
16	ILLINOIS	165,000	41	NEW JERSEY	24,000
17	TENNESSEE	159,000	42	NEVADA	22,500
18	VERMONT	158,000	43	WEST VIRGINIA	22,000
19	INDIANA	145,000	44	MONTANA	21,000
20	VIRGINIA	130,000	45	NEW HAMPSHIRE	20,000
21	SOUTH DAKOTA	120,000	46	HAWAII	10,700
22	ARIZONA	116,000	47	DELAWARE	10,000
23	GEORGIA	102,000	48	WYOMING	7,000
24	OREGON	100,000	49	RHODE ISLAND	2,200
25	OKLAHOMA	99,000	50	ALASKA	700
				TOTAL	APP. 9,525,000

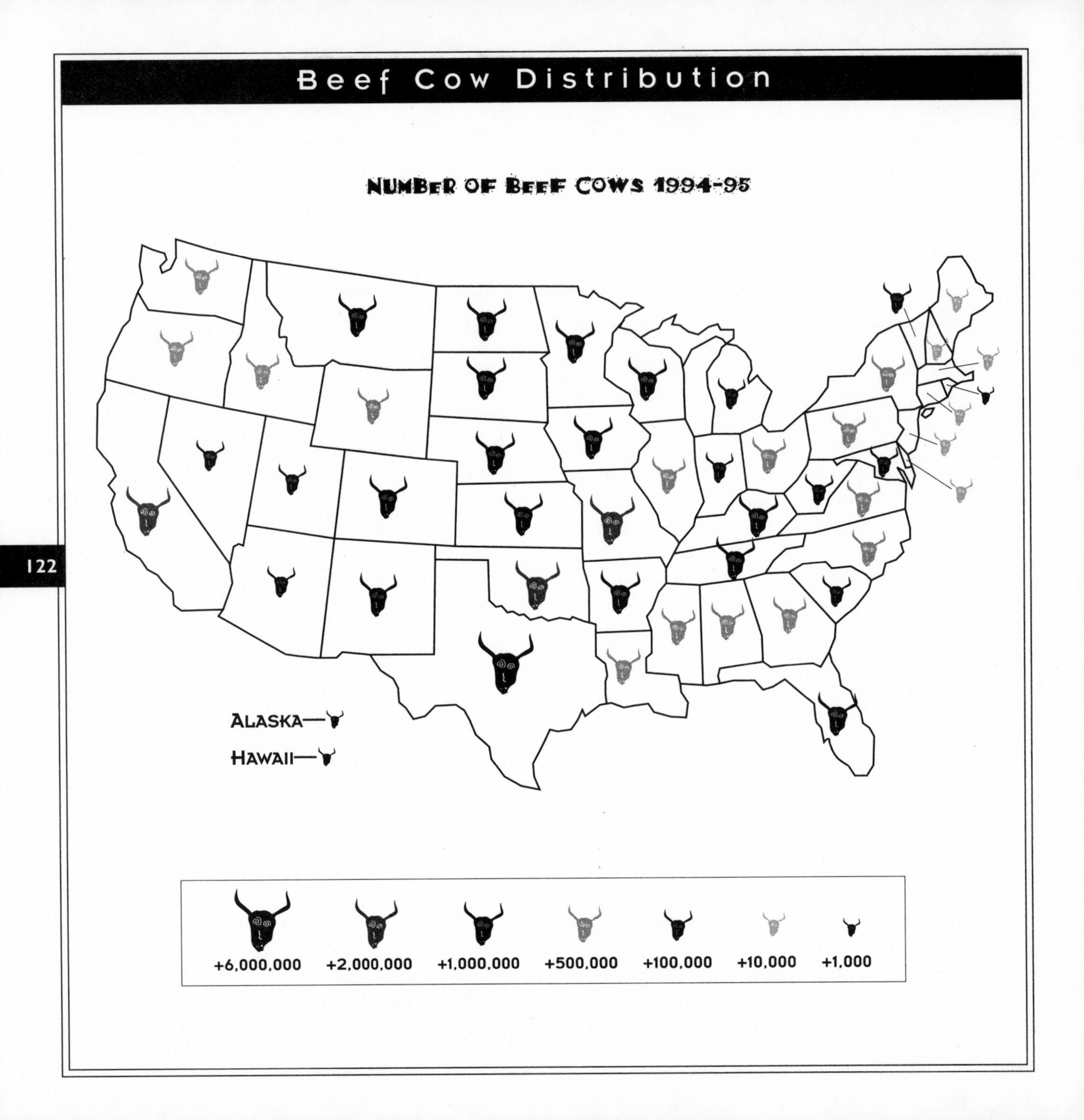

Beef Cow Distribution
NUMBER OF BEEF COWS 1994-95
ALASKA
HAWAII
+6,000,000
+2,000,000
+1,000,000
+500,000
+100,000
+10,000
+1,000

Ranking of Beef States

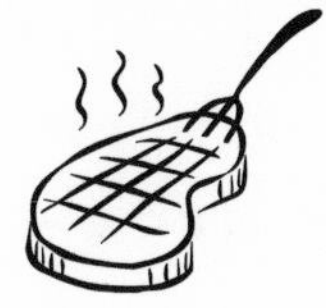

The ranking of states in terms of number of beef cows. This does not include calves, steers, or bulls.

NATIONAL AGRICULTURAL STATISTICS SERVICE 1994-95

RANK	STATE	BEEF COWS
1	TEXAS	6,600,000
2	MISSOURI	2,350,000
3	OKLAHOMA	2,200,000
4	CALIFORNIA	2,140,000
5	NEBRASKA	1,960,000
6	SOUTH DAKOTA	1,800,000
7	WISCONSIN	1,690,000
8	KANSAS	1,650,000
9	MONTANA	1,610,000
10	IOWA	1,430,000
11	KENTUCKY	1,330,000
12	FLORIDA	1,300,000
13	TENNESSEE	1,290,000
14	ARKANSAS	1,080,000
15	MINNESOTA	1,040,000
16	NORTH DAKOTA	1,010,000
17	ALABAMA	970,000
18	COLORADO	900,000
19	VIRGINIA	870,000
20	PENNSYLVANIA	810,000
21	GEORGIA	810,000
22	NEW YORK	780,000
23	MISSISSIPPI	740,000
24	WYOMING	740,000
25	IDAHO	740,000

RANK	STATE	BEEF COWS
26	NEW MEXICO	730,000
27	OREGON	720,000
28	ILLINOIS	700,000
29	OHIO	640,000
30	LOUISIANA	610,000
31	NORTH CAROLINA	590,000
32	WASHINGTON	580,000
33	INDIANA	470,000
34	MICHIGAN	465,000
35	UTAH	430,000
36	ARIZONA	365,000
37	SOUTH CAROLINA	270,000
38	WEST VIRGINIA	270,000
39	NEVADA	265,000
40	VERMONT	172,000
41	MARYLAND	160,000
42	HAWAII	92,000
43	MAINE	57,000
44	CONNECTICUT	40,000
45	MASSACHUSETTS	37,000
46	NEW JERSEY	36,000
47	NEW HAMPSHIRE	24,000
48	DELAWARE	13,000
49	ALASKA	3,700
50	RHODE ISLAND	3,400
	TOTAL	APP. 45,583,100

Ranking of Dairy Countries

The ranking of countries for the number of milk cows and milk production.

NATIONAL AGRICULTURAL STATISTICS SERVICE 1991

RANK	COUNTRY	MILK COWS	MILK IN TONS
1	RUSSIA	41,481,000	101,720,000
2	INDIA	31,000,000	27,000,000
3	BRAZIL	15,500,000	13,800,000
4	USA	9,990,000	67,370,000
5	MEXICO	6,440,000	10,200,000
6	GERMANY	6,300,000	30,500,000
7	FRANCE	5,400,000	25,880,000
8	POLAND	4,707,000	14,906,000
9	U.K.	3,206,000	14,710,000
10	ITALY	2,800,000	4,626,000

Ranking of Cow Countries

The ranking of countries for the total number of cattle.

NATIONAL AGRICULTURAL STATISTICS SERVICE 1992

RANK	COUNTRY	CATTLE
1	INDIA	271,437,000
2	BRAZIL	130,700,000
3	RUSSIA & REPUBLICS	112,400,000
4	CHINA	104,214,000
5	UNITED STATES	100,110,000
6	ARGENTINA	50,029,000
7	AUSTRALIA	25,075,000
8	FRANCE	19,926,000
9	GERMANY	17,134,000
10	COLOMBIA	16,145,000

Major Cattle Trails 1866-1895

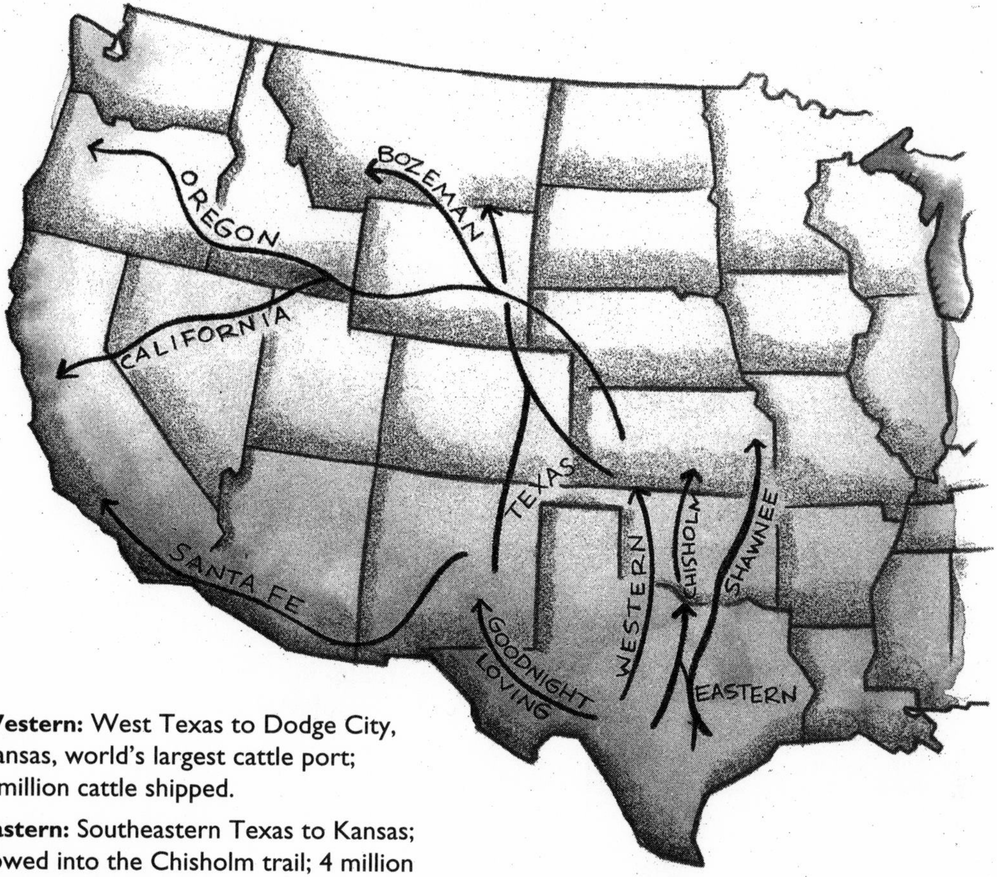

1. **Western:** West Texas to Dodge City, Kansas, world's largest cattle port; 5 million cattle shipped.
2. **Eastern:** Southeastern Texas to Kansas; flowed into the Chisholm trail; 4 million cattle shipped.
3. **Chisholm:** Wichita, Kansas to Anadarko, Oklahoma; connected with the Eastern trail, but it never entered Texas.
4. **Goodnight-Loving:** Southwest Texas to Fort Sumner, New Mexico; 1/4 million cattle moved.
5. **Texas:** Offshoot of the Goodnight-Loving and Western trails going up to Montana.
6. **Bozeman:** Nebraska to Bozeman, Montana.
7. **Oregon:** Nebraska to Oregon and Idaho.
8. **Santa Fe:** Santa Fe, New Mexico to Southern California.
9. **California:** Idaho to Northern California.
10. **Shawnee:** East Texas to Missouri.

Cow Products

A 1,000-pound steer produces approximately 430 pounds of beef; around 1 percent of the gross weight is lost during processing, and the remaining 570 pounds might go into:

- bandages, baseball gloves, buttons
- "camel" hair brushes, candles, ceramics, chewing gum, combs cosmetics, crayons
- deodorants, detergent
- emery boards, explosives
- floor wax, fly swatters
- glue
- hood ornaments, hydraulic brake fluid
- ice cream, imitation tortoise shell, insecticides, insulation, insulin, linoleum, luggage
- margarine, marshmallows, mayonnaise, medical sutures
- natural sausage casing, neatsfoot oil
- paints, perfumes, piano keys, photographic film, phonograph records, plastics
- runway foam for airports
- shaving cream, sheetrock, soaps
- textiles, trypsin
- wallets, wallpaper
- and yogurt thickener.

Cheese Fact:

The holes in Swiss Cheese originate from trapped carbon dioxide during the fermentation process.

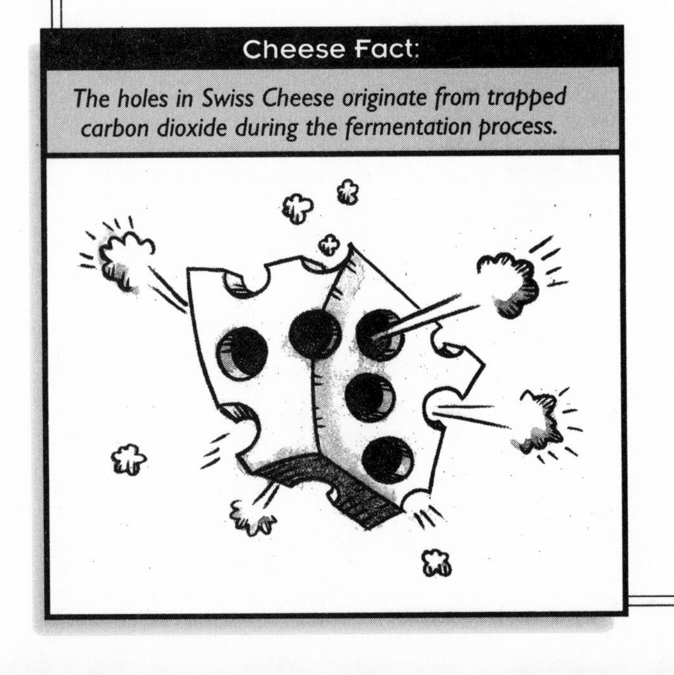

Cheeses of America

- **AMERICAN**—Cheese processed from other cheeses.
- **AMERICAN BLUE**—American-made Blue cheese.
- **AMERICAN CHEDDAR**—America's number one bestseller.
- **BANDON**—Oregon Cheddar made from Jersey milk.
- **BELMONT**—Wisconsin-made Brie.
- **BRICK**—A soft, mild cheese.
- **BURMEISTER**—Delicious stuff that has a German-sounding name.
- **CAMOSUN**—Semisoft, Gouda-like cheese.
- **COLBY**—From Colby, Wisconsin. An open-textured Cheddar.
- **COON**—Cheddar variety named after the raccoon.
- **CORNHUSKER**—Cheddar variety from Nebraska.
- **COTTAGE CHEESE**—Curd from sour milk, cooked and washed.
- **CREAM CHEESE**—Cheese that is fortified with cream and not fermented.
- **CREOLE**—Louisiana cheese.
- **FARMERS CHEESE**—Cheese that is not aged.
- **HAND**—A strong flavored cheese.
- **HERKIMER**—New York Cheddar variety.
- **LIEDERKRANZ**—Limburger variety from Ohio.
- **MONTEREY OR MONTEREY JACK**—From Monterey, California and first made by David Jack.
- **MYSOST**—A cheese popular in the midwest states.
- **OLD HEIDELBERG**—Limburger variety from Illinois.
- **PINEAPPLE**—Cheddar placed in a net and shaped like a pineapple. First made in Connecticut during the Art Deco craze of the middle 1800s.
- **POONA**—Mild Limburger from New York.
- **SAGE**—Spiced Cheddar variety.
- **SWISS**—The cheese with holes; second only to Cheddar in popularity.

Famous Cows

- **Achelous**—A Greek river god who turned himself into a bull.
- **Apis or Hap**—Bull god of the sun from Memphis, Egypt.
- **Audhumla**—Norse cow god.
- **Babe**—Paul Bunyan's Blue Ox; seven ax handles fit between her eyes and she weighed more than the combined weight of all the fish that got away.
- **Beecher Alinda Ellen**—The record milking cow, giving 5,392.7 gallons in a year; a Holstein.
- **Bevo I**—The first University of Texas mascot, in 1916; irate A&M Aggie fans served the steer up at a banquet.
- **Black Diamond**—A Bison and the model for the Buffalo nickel.
- **Brown Eyes**—The Jersey cow in Buster Keaton's 1925 film "Go West."
- **Donnetto**—4,300-pound record cow; a Chianina.
- **Elm Farm Ollie**—In 1930, the first cow to fly in an airplane; she was also milked while flying.
- **Enlil**—Sumerian bull god of storms.
- **Enola Gay**—The cow of Captain Paul Tibbets' mother and the name of the plane over Hiroshima.
- **Ferdinand**—In a story by Munro Leaf, the name of a Spanish fighting bull who would rather smell flowers than fight.
- **Geush Uravn**—Soul of bovines in Persian mythology.
- **Hathor**—Sacred cow of Egypt.
- **Islero**—The bull that gored and killed the world's greatest bullfighter while the matador was giving the final fatal blow to the bull.
- **L&S Telstar Sally May Twin**—Butterfat record of 6.2%; equal to 2,067 pounds. A Brown Swiss.
- **Minotaur**—Mythical half man/half bull (taur from taurus).
- **Mrs. O'Leary's cow**—The Great Chicago Fire of 1871 started in the O'Leary's barn, but there was never any proof that her cow really started that fire.
- **Ninlil**—Sumerian lunar cow god.
- **Norman**—Billy Crystal's favorite cow in the movie *City Slickers*.
- **Old Ben**—A Shorthorn and Hereford cross that weighed 4,720 pounds; 16'2" from nose to tip of tail; girth of 13'8"; height of 6'4".
- **You'll Do, Lobelia**—The original model for Elsie, the Jersey on the Borden Milk can.

THE FATTEST COW

Durham Ox, a Shorthorn in the 19th century that had layers of fat on it.

THE TALLEST COW

Chianina, modeled for Michelangelo.

THE COW WITH THE MOST MILK

Holstein, the nation's number-one dairy cow.

THE SMALLEST COW

Dexter, bred a small size for household living.

Corriente, used in rodeo for roping contests.

THE COW WITH THE RICHEST, BEST-TASTING MILK

Jersey, also great for making ice cream.

Cow Crosses

- **American Breed** = $\frac{1}{2}$ Zebu; $\frac{1}{8}$ each of Charolais, Hereford, Shorthorn, Bison
- **Amerifax** = Angus x Bison
- **Ankina** = Angus x Chianina
- **Black Baldy** = Angus x Hereford
- **Brahorn** = Brahman x Shorthorn
- **Bralers** = Brahman x Salers
- **Bravon** - Brahman x Devon
- **Charbray** = Charolais x Brahman
- **Charford** = Charolais x Hereford
- **CharSwiss** = Charolais x Brown Swiss
- **Chiangus** = Chianina x Angus
- **Chiford** = Chianina x Hereford
- **Hash Cross** = Charolais, Hereford, Lincoln Red
- **Hays Converter** = American Brown Swiss, Hereford, Holstein
- **Holgus** = Holstein x Angus
- **Jamaican Hope** = Jersey x Zebu
- **Jamaican Red** = South Devon, Red Poll, and Zebu
- **Makaweli** = From Hawaii; Shorthorn x Devon
- **PeeWee** = From Canada; Angus, Charolais, Hereford, and Galloway cross.
- **Regus** = Red Angus graded up from a Hereford
- **Sabre** = $\frac{7}{8}$ Sussex x $\frac{1}{8}$ Brahman

Cow Fact:

A cow's average temperature is 101.5 degrees Fahrenheit.

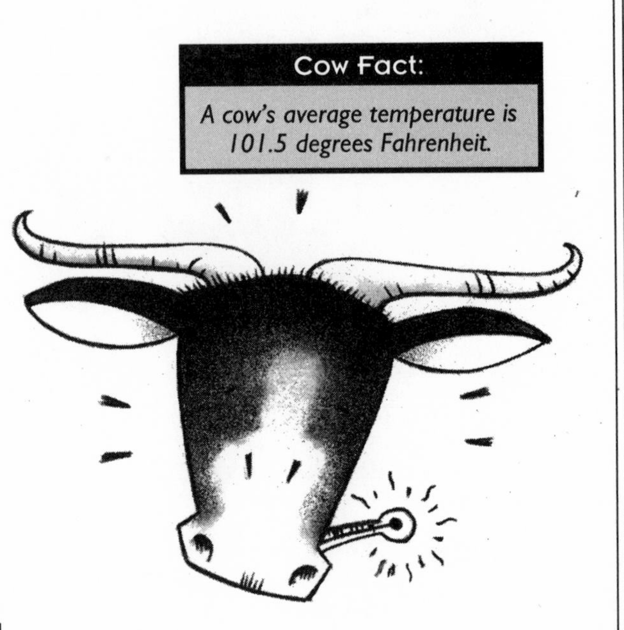

- **Senepol** = from the Caribbean; N'Dama x Red Poll
- **Simbrah** = Simmental x Brahman
- **Simmalo** = Simmental x Bison
- **Victoria** = $^{3}/_{4}$ Hereford, $^{1}/_{4}$ Brahman
- **Ranger** - A conglomerated cattle breed developed for the open range. Two different herds started during the 1950s: the Wyoming herd consisted of Simmental, Hash Cross, Beefmaster, Hereford, Brahman, and Highland x Shorthorn bulls; and the California herd consisted of American Brown Swiss, Red Angus, Red Holstein, Beefmaster, and Hash Cross x Hereford cows.

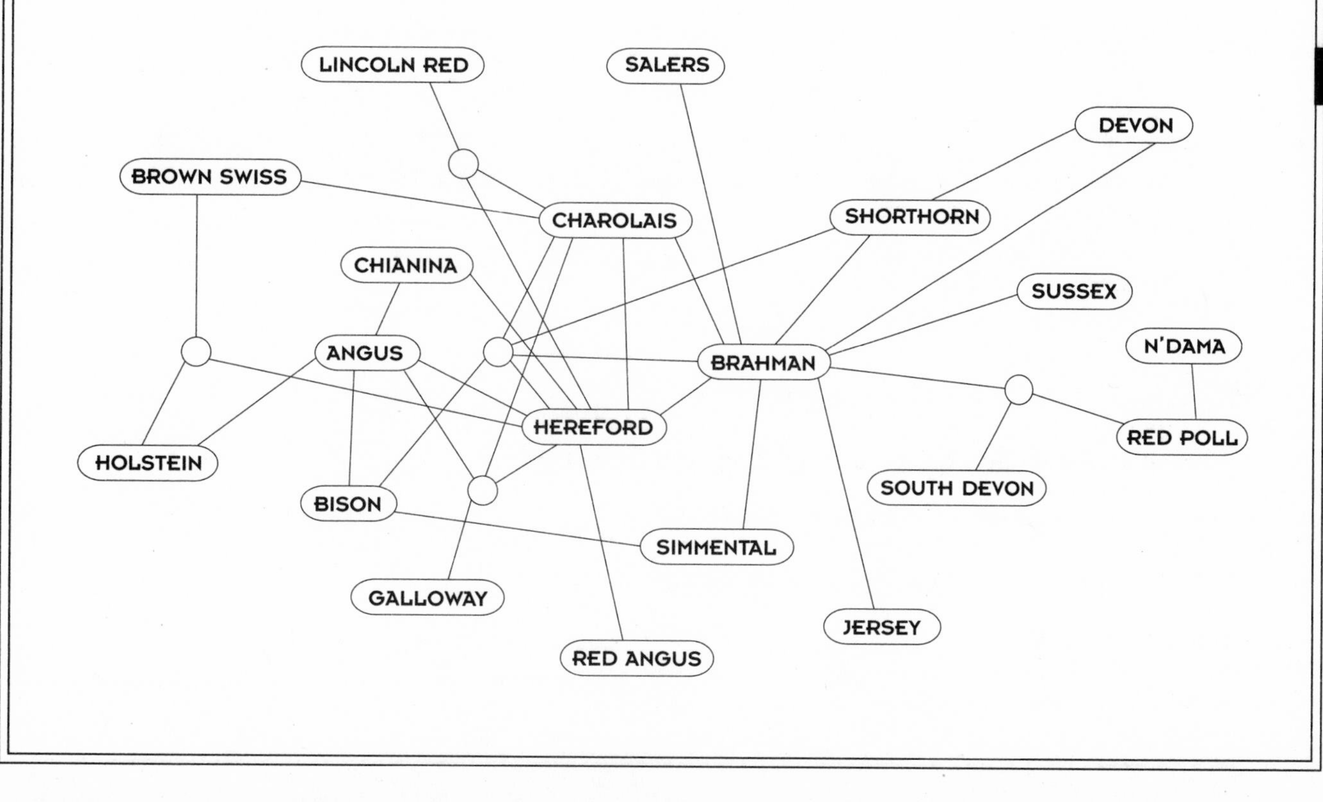

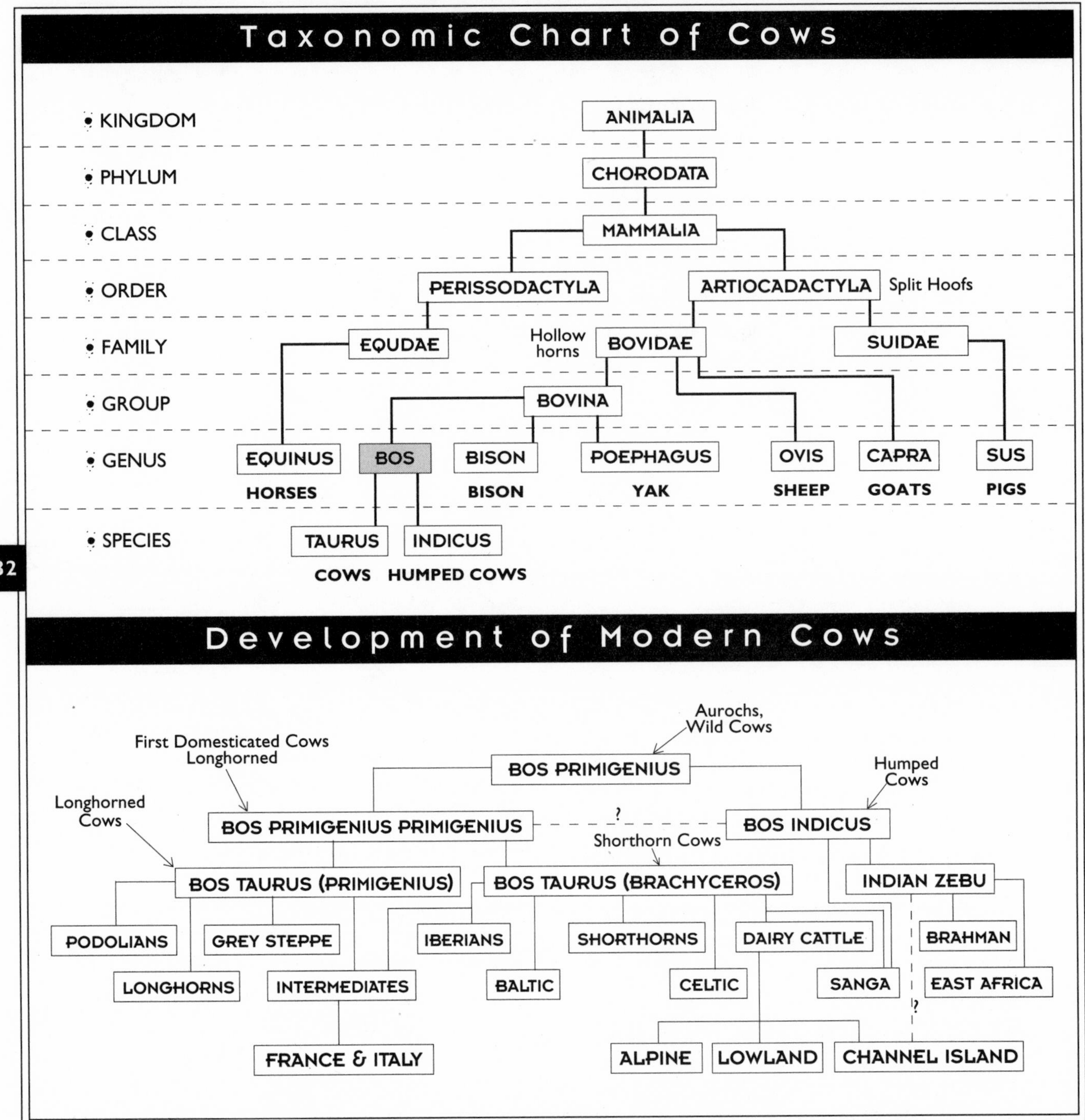
Taxonomic Chart of Cows
KINGDOM
PHYLUM
CLASS
ORDER
FAMILY
GROUP
GENUS
SPECIES
ANIMALIA
CHORODATA
MAMMALIA
PERISSODACTYLA
ARTIOCADACTYLA
Split Hoofs
EQUDAE
Hollow horns
BOVIDAE
SUIDAE
BOVINA
EQUINUS
HORSES
BOS
BISON
BISON
POEPHAGUS
YAK
OVIS
SHEEP
CAPRA
GOATS
SUS
PIGS
TAURUS
COWS
INDICUS
HUMPED COWS
Development of Modern Cows
Aurochs, Wild Cows
First Domesticated Cows Longhorned
BOS PRIMIGENIUS
Humped Cows
Longhorned Cows
BOS PRIMIGENIUS PRIMIGENIUS
?
BOS INDICUS
Shorthorn Cows
BOS TAURUS (PRIMIGENIUS)
BOS TAURUS (BRACHYCEROS)
INDIAN ZEBU
PODOLIANS
GREY STEPPE
IBERIANS
SHORTHORNS
DAIRY CATTLE
BRAHMAN
LONGHORNS
INTERMEDIATES
BALTIC
CELTIC
SANGA
EAST AFRICA
?
FRANCE & ITALY
ALPINE
LOWLAND
CHANNEL ISLAND

Glossary

Angular—Wedgy body shape in some dairy cattle.

Aurochs—Urus. The original wild cattle of Europe, now extinct.

Backcross—The mating of an f1 animal to one of its parents.

Bay—Reddish brown color.

Baltic Red—Type of Brachyceros found in northern Europe.

Bloat—Gas caught in the rumen; noticeable on the cow's upper left side.

Blocky—Deep, wide, low-set body; common shape of beef cattle.

Bolus—Large pill or regurgitated food mass ready to be swallowed.

Brachyceros—Means "short horns"; the forehead is wider than it is long. The cow usually has high withers and rump, giving a dipped back. Probably developed or domesticated from the primigenius. Source of all dairy cattle. Same as longifrons.

Bred—Animal that is pregnant.

Breed—Genetically pure line having similar conformation and able to produce like offspring.

Breed Characteristic—Specific variations: color, neatness, refinement.

Breed Society—A group organized for the improvement and promotion of a specific breed. Also the group that usually publishes the Herdbook.

Brindle—Streak or spots on a gray or brownish background.

Brockled-Face—Spotted face.

Buck-Kneed or Knee-Sprung—Knees too far forward.

Buff—Yellow-reddish color or a light skin color.

Bull—Uncastrated male.

Bulling Cow—Cow in heat.

Bulldog—A genetic condition occasionally seen in Dexter calves.

Butter—Churned cream or milk that coagulates.

Calf-Hocked—Knees too far back on a calf.

Calf—Sexually immature young cow.

Carcass—The dressed body of a meat animal.

Carcass Yield—Carcass weight divided by live weight.

Celtic—A type of Brachyceros cattle. Related to Iberian and Baltic Reds.

Cervical—The neck region.

Cheese—Normally, fermented curd, except Ricotta, which is made from whey.

Closebreeding—Mating closely related animals.

Colostrum—Milk that is secreted by a mother cow the first few days after birth; high in antibodies and laxatives.

Commercial Grade—Low grade quality of meat.

Concentrates—Processed feed high in energy, low in crude fiber.

Conformation—Body shape; helps distinguish beef, dairy, draft, or multipurpose breeds.

Constitution—Chest capacity; or, adaptability to the environment.

Cow—Mature female with at least one calf.

Cow/Calf Operation—Breeding of cows for the raising and selling of calves.

Cow-Hocked—Hocks bent inward, toes outward.

Creep—Feeding area only for calves.

Cream—The good stuff in milk, made up of mostly butterfat and protein.

Crest—Top part of the neck.

Criollo—Spanish for "native"; used for any indigenous cattle.

Crossbred—Dam and sire are of different purebred breeds.

Cud—A bolus of regurgitated food.

Curds—The thick part of coagulated milk.

Cutability—Estimated percentage of meat for the carcass weight.

Cutter Grade—The lowest cut of meat.

Dam—Female parent.

Disposition—Temperament of the animal.

Dogie—A motherless calf.

Dominant—The appearing genetic trait; may hide a recessive trait.

Double-Muscling—A condition seen in some beef breeds: actually longer muscles.

Draft—An animal for pulling a plow or cart.

Dressing Percent—Percent yield of carcass to live animal.

Drover—Trail rider.

Dry Cow—Cow that is not producing milk.

Dual Purpose—A cow breed raised for two uses, that is, either beef and draft or beef and dairy; hardly ever dairy and draft.

Dun—Red to yellow in color or a dull grayish brown.

Extensive—An operation requiring low cost or effort to get production (e.g., a rangeland operation).

f1—First generation of a cross between a purebred bull and cow.

Family—Lineage of an animal as traced through either male or female.

Fawn—Brown or red-yellow in color.

Feeder—A calf that will gain weight if placed on feed; usually goes to market.

Feedlot—Place to feed and finish cattle for market.

Feral—Domesticated animal that went back to a wild state.

Finching—Colored stripe down the center of the back.

Finish—To fatten an animal for slaughter. Also, degree of fatness.

FNOCL—The French dairy herd improvement program.

Frame—Size and conformation.

Freshen—Said of a cow that is ready to give birth.

Freemartin—The female of a set of twins with the other calf a male. Nine-tenths are infertile.

Glum—Social group, as seen in the Watusi.

Gomer—Out of control animal.

Grade—An animal having unregisterable parents. Or, how well a cow fulfills a class.

Grading Up—Crossing purebred sires with graded animals.

Hand—Four inches; common measurement system for horses and cattle.

Hard Keeper—Unthrifty; grows or fattens slowly on any type of feed.

Hay—Dry forage.

Heifer—A young female cow, especially one that has never had a calf.

Herdbook—The publication of a breed's ancestry and characteristics. Sort of the breed's family tree.

Heterosis—When a crossbreed outproduces the average of the parents.

Heterozygous—The trait where genes for two differing characteristics are present; e.g., a Black Angus having a recessive gene for red hair would be a heterozygote for that specific trait.

Homozygous—The trait where genes for only one allelomorphic characteristic are present; e.g., a Red Angus would be homozygous for red hair color.

Husbandry—The raising of livestock.

Hybrid Vigor or Heterosis—When a crossbreed outproduces the average of the parents.

Iberian—Type of Brachyceros from the eastern and southern Mediterranean.

Intensive—High effort or cost to get production; e.g., a feedlot operation.

Lactation—abbreviated Lac—Length of time giving milk.

Lactose—Whole milk's major carbohydrate at 4.9 percent.

Land Race—Livestock for which both natural and human selection play a role in development.

Leanness—Amount of fat.

Leggy—Long in the leg.

Line—Having a close relationship to a specific animal or a set of animals.

Linebreeding—Breeding to keep certain favored characteristics.

Linecross—Cross of two different lines.

Longifrons—The forehead is wider than it is long. Same as brachyceros.

Marbling—Fat in the muscle tissue.

Maturity—Age of the cattle.

Muley—Hornless cow.

NCDHIP—National Council of Dairy Herd Improvement Program.

Nicking—When an offspring's characteristics exceed that of its parents.

Outcross—Cross of the same breed, but unrelated cows.

Overmarking—Small black spots that appear on otherwise all-white cattle.

Oxen—Working cattle, male or female.

Pasture—Grassland.

PATLQ—Canadian equivalent of the Dairy Herd Improvement Program.

Piebald—Speckled black and white.

Podolian—An old cattle breed still seen on the steppes of Russia and with some of the Italian breeds; subgroup of primigenius.

Points—The ears, nose, hoofs, and switch.

Poll—The space between the ears on top of the head.

Polled—Naturally hornless or with the horns removed.

Primigenius—The first cattle, from which all cattle are thought to be descended. Characterized by long horns. Podolians are considered a subgroup. Also see Aurochs.

Purebred—Registered animal or animal eligible for registration in a herdbook.

Quality—Refinement of an animal: neat, well-chiseled head, fine hair, clean bone.

Quality Yield—Estimated desirability of beef, considering marbling and animal's age.

Range—Extensive area of grassland where cattle wander and graze.

Rangy—Long in the body.

Rbgh—Recombinant bovine growth hormone; used to increase milk production in cows.

Recessive—A hidden genetic trait that only appears when matched with the same recessive trait; also see dominant.

Roan—Intermixture of red and white hairs, not considered spotted.

Roughage—Grass or hay, rather than concentrates.

Rumen—First stomach.

Ruminate—To chew cud.

Scrub—Odd mixture of grade cattle.

Scurs—Buttons; small bits of horn tissue attached to skin on polled animals.

Sebum—Skin secretions.

Shire—English for county.

Sheath—Skin hanging underneath the body.

Sire—Male parent.

Skewbald—Speckled white and red or other color, but not black.

Springers—Jumping calves in the springtime.

Stag—Male castrated after secondary sexual characteristic develops.

Steer—Male castrated before secondary sexual characteristic develops.

Stocker Cattle—From weaning to 800 pounds; feeders.

Stover—Dried stalks and leaves, but not grain of corn or milo.

Straw—Plant residue after threshing removes seeds.

Suckler—A cow that is giving milk to a calf.

Subcutaneous—Under the skin.

Symmetry—Balanced development of all the parts.

Tandem Selection—Breeding for one trait at a time.

Terminal Breed—All offspring go to market.

Thrifty—Healthy and vigorous in appearance. Opposite of a hard keeper.

Thoracic—The chest region of cattle.

Triple Purpose—A cow breed raised for dairy, draft, and beef uses.

Type—Ideal of perfection, combining all characteristics for a breed.

Upgrading—Breeding to get purebred cattle from grade cattle.

Urus—Aurochs. The original wild cattle of Europe, now extinct.

Utility Grade—A low quality grade of meat.

Variety—Like subspecies, but less defined.

Veal—Meat from a calf less than three months old, strictly fed milk.

Wasty—A carcass with too much fat. Also, paunchy live animals.

Weaning—The end of suckling.

Weanling—A weaned calf.

Whey—The watery part of coagulated milk.

Withers—The top of the shoulder.

Yearling—Twelve to 18 months of age.

Zebu—A *Bos indicus* type of cattle (e.g., Brahman).

Selected References

Alderson, G. Lawrence. *The Chance to Survive.* London: A. & C. Black Pub. Ltd., 1990.

The American Livestock Breeds Conservancy. *Taking Stock: The North American Livestock Census.* Blacksburg: McDonald & Woodward Pub., 1994.

Biggle, Jacob. *Biggle Cow Book.* Philadelphia: W. Atkinson Co., 1898.

Diffloth, Paul. *Bovides.* Paris: J.B. Bailliere., 1904.

Ensminger, M. Eugene. *Dairy Cattle Science.* 3rd ed. Danville: Interstate Pub., 1993.

Ensminger, M. Eugene. *Beef Cattle Science.* 6th ed. Danville: Interstate Pub., 1978.

Felius, Marleen. *Genus Bos: Cattle Breeds of the World.* Rahway: MSD AGVET, 1985.

Friend, John B. *Cattle of the World.* Poole: Blandford Press, 1978.

Gallant, Marc. *The Cow Book.* New York: Knopf, 1983.

Jordan, Terry G. *North American Cattle-Ranching Frontiers: Origins, Diffusion, and Differentiation.* Albuquerque: University of New Mexico Press, 1993.

Jones, Evans. *The Word of Cheese.* New York: Knopf, 1976.

Mason, Ian L. *A World Dictionary of Livestock Breeds, Types and Varieties.* Wallingford: C.A.B. International, 1988.

Porter, Valerie. *Cattle. A Handbook to the Breeds of the World.* London: Christopher Helm, 1991.

Rouse, John E. *World Cattle* 3 vol. Norman: University of Oklahoma Press, 1970-1974.

Sambraus, Hans H. *A Colour Atlas of Livestock Breeds.* London: Wolfe Pub., 1992.

Webb, Byron H. *Fundamentals of Dairy Chemistry.* 2nd ed., Westport: AVI Pub. Co., 1974

Cow Fact:

Cows can see color.

Breeders Associations

American Angus Association
3201 Frederick Boulevard
St. Joseph, Missouri 64506

Ayrshire Breeder's Association
PO Box 1608
Battleboro, Vermont 05302-1608

Barzona Breeders Association
of America
PO Box 631
Prescott, Arizona 86302

Beefmaster Breeders Universal
6800 Park Ten Blvd.,
Suite 290 West
San Antonio, Texas 78213

American Blonde d'Aquitaine
Association
PO Box 12341
Kansas City, Missouri 64116

United Braford Breeders
422 East Main Street, Suite 218
Nacogdoches, Texas 75961

American Brahman Breeders
Association
1313 La Coucha Lane
Houston, Texas 77054-1890

American Bralers Association
HCR 61, Box 41
Ganado, Texas 77962

Brangus Breeders Association
PO Box 696020
San Antonio, Texas 78269-6020

Brown Swiss Cattle Association,
USA
Box 1038
Beloit, Wisconsin 53511-1038

Societe Des Eleveurs de Bovins
Canadiens
468 Rue Dolbeau
Sherbrooke, Quebec J1G 2Z7
Canada

American-International Charolais
Association
PO Box 20247
Kansas City, Missouri 64195

American Chianina Association
PO Box 890
Platte City, Missouri 64079

North American Corriente
Association
PO Box 12359
North Kansas City, Missouri
64116

Devon Cattle Association, Inc.
RR 1, PO Box 93
Bunkie, Louisiana 71322-9709

American Dexter Cattle
Association
Route 1, PO Box 378
Concordia, Missouri 64020

Dutch Belted Association
of America
c/o American Livestock Breeds
Conservancy
PO Box 477
Pitsboro, North Carolina 27312

American Galloway Breeder's
Association
c/o Bob Mullendore
310 West Spruce
Missoula, Montana 59802

American Gelbvieh Association
10900 Dover St.
Westminster, Colorado 80021

American Guernsey Association
7614 Slate Ridge Road
PO Box 666
Reynoldsburg, Ohio 43068-0666

American Hereford Association
1501 Wyandotte
PO Box 014059
Kansas City, Missouri 64101

American Highland Cattle
Association
Livestock Exchange Building,
Suite 200
4701 Marin Street
Denver, Colorado 80216

Holstein Association 1
Holstein Place
Battleboro, Vermont 05302-0808

American Jersey Cattle Club
6486 East Main Street
Reynoldsburg, Ohio 43068-2362

North American Limousin
Foundation
7383 South Alton Way
PO Box 4467
Englewood, Colorado 80155

American Maine-Anjou
Association
528 Livestock Exchange Building
Kansas City, Missouri 64102

Marky Cattle Association
Box 198
Walton, Kansas 67151-0198

Canadian Meuse-Rhine-Ijssel Association
Box 235
Claresholm, Alberta T0L 0T0
Canada

American Milking Devon Association
c/o Sue Randall
Old Bay Road
New Durham, New Hampshire 03855

American Milking Shorthorn Society
PO Box 449
Beloit, Wisconsin 53512-0449

American Murray-Grey Association
PO Box 34590
North Kansas City, Missouri 64116

North American Normande Association
11538 Spudville Road
Hibbing, Minnesota 55746

Piedmontese Association of the U.S.
Livestock Exchange Building, Suite 108
Denver, Colorado 80216

American Pinzgauer Association
21555 State Road 698
Jenera, Ohio 45841

Red Angus Association of America
Box 4201 I-35
North Denton, Texas 76207-7443

American Red Polled Association
PO Box 35519
Louisville, Kentucky 40232

American Romagnola Association
PO Box 450
Navasota, Texas 77868-0450

American Salers Association
5600 South Quebec, Suite 220A
Englewood, Colorado 80111-2207

Santa Gertrudis Breeders International
PO Box 1257
Kingsville, Texas 78364

American Senepol Association
PO Box 901594
Kansas City, Missouri 64190-1594

American Shorthorn Association
8288 Hascall Street
Omaha, Nebraska 68124

American Simmental Association
1 Simmental Way
Bozeman, Montana 59715

North American South Devon Association
PO Box 68
Lynnville, Iowa 50153

Sussex Cattle Association of America
PO Box 107
Refugio, Texas 78377

American Tarentaise Association
PO Box 34705
North Kansas City, Missouri 64116

Texas Longhorn Breeders Association of America
PO Box 4430
Fort Worth, Texas 76106

World Watusi Association
PO Box 230
Crawford, Colorado 81415

Welsh Black Association, USA
RR 1 Box 76B
Shelburn, Indiana 47879

North American White Park Cattle Association
HC 87 Box 2214
Big Timber, Montana 59011

Cow Observers Worldwide (COW)
c/o Cowtree Collector
240 Wahl Ave.
Evans City, Pennsylvania 16033

A Cow Watcher's Life List

- ❑ Africander
- ❑ American Breed
- ❑ Amerifax
- ❑ Angus
- ❑ Ankina
- ❑ Aurochs or Urus - see glossary
- ❑ Ayrshire
- ❑ Barzona
- ❑ Beef Friesian
- ❑ Beefalo
- ❑ Beefmaster
- ❑ Beligian Blue
- ❑ Belgian Red Pied
- ❑ Belties - Belted Galloway
- ❑ Bison
- ❑ Black Baldy
- ❑ Blonde d'Aquitaine
- ❑ Bra-swiss
- ❑ Braford
- ❑ Brahman
- ❑ Brahmousin
- ❑ Brahorn
- ❑ Bralers
- ❑ Brangus
- ❑ Bravon
- ❑ British White
- ❑ Brown Swiss
- ❑ Canadienne
- ❑ Charbray
- ❑ Charford
- ❑ Charolais
- ❑ Charswiss
- ❑ Charwiss
- ❑ Chiangus
- ❑ Chianina
- ❑ Chiford
- ❑ Corriente
- ❑ Danish Red
- ❑ Devon
- ❑ Dexter
- ❑ Dutch Belted
- ❑ Fjallrus
- ❑ Galloway
- ❑ Gelbvieh
- ❑ Gir
- ❑ Grey Steppe
- ❑ Guernsey
- ❑ Hash Cross
- ❑ Hays Converter
- ❑ Hereford
- ❑ Highland
- ❑ Holgus
- ❑ Holstein
- ❑ Illawara
- ❑ Jamaican Hope
- ❑ Jamaica Red
- ❑ Jersey
- ❑ Kerry
- ❑ Latvian Red
- ❑ Limousin
- ❑ Lincoln Red
- ❑ Lineback, Randall
- ❑ Makaweli
- ❑ Maine-Anjou
- ❑ Marchigiana
- ❑ Meuse Rhine Yssel (MRY)
- ❑ Milking Shorthorn
- ❑ Murray-Grey
- ❑ Nellore
- ❑ Normande
- ❑ Norwegian Red
- ❑ Pee Wee
- ❑ Piedmont
- ❑ Pinzgauer
- ❑ Ranger
- ❑ Red and White Holstein
- ❑ Red Angus
- ❑ Red Poll
- ❑ Regus
- ❑ Romagnola
- ❑ Sabre
- ❑ Salers
- ❑ Santa Gertrudis
- ❑ Senepol
- ❑ Shorthorn
- ❑ Simbrah
- ❑ Simmalo
- ❑ Simmental
- ❑ South Devon
- ❑ Sussex
- ❑ Tarentaise
- ❑ Texas Longhorn
- ❑ Victoria
- ❑ Watusi
- ❑ Welsh Black
- ❑ White Park
- ❑ Zebu, Indo-Brazilian

FALCON PRESS ANNOUNCES THE FIRST EVER:
A FIELD GUIDE TO COWS
CHRISTMAS COW COUNT
(AND CONTEST)

Cow connoisseurs from everywhere are invited to participate in this one-of-a-kind event, very loosely modeled after nationwide Christmas bird counts. Participants need simply to find some cows, identify them, record the sightings, complete the attached entry form, and mail it to Falcon Press. Prizes will be awarded by random drawing. If it works, maybe we'll make it an annual event.

Christmas Cow Count Rules and Other Fine Print

1. Cows seen on those Christmas bird counts are valid for this contest, but not vice versa.
2. All cows must be live.
3. All cows must be seen. Cow sounds don't count.
4. Unidentifiable breeds should be counted as "unknown," "unclear," "wierd," "strange," or "mutt."
5. Cow and calf pairs only count as one. Calves already weaned and on their own count as one.
6. Counting must be conducted on a single day, and only between sunrise and sunset (or sunup and sundown, as real cowboys say). Any day between December 15 and December 30, 1996, is okay.
7. Counts may be conducted by individuals, pairs, families, or teams. (You figure out how to divide up the prize if you win.)
8. Contestants should use the handy entry form in *A Field Guide To Cows,* but any written entry is okay. However, entry forms with traces of cow manure will be disqualified, regardless of authenticity or numbers of cows seen.
9. Completed entry forms must be mailed to Christmas Cow Count, Falcon Press, P.O. Box 1718, Helena, Montana 59624 and be postmarked no later than January 15, 1997. (No faxes or phone calls will be accepted.)
10. *A Field Guide To Cows* author John Pukite will draw the names of at least ten winners out of his Mallard Seed cap sometime in the spring (after calving and before branding). Winners will be notified by mail.
11. Winners will receive a cow prize of some type—maybe a set of cow drinking mugs (for people, not cows), cow canisters, another copy of this book, beef jerky, cheese, seed/feed/breed hats, and other stuff we might think of or want to get rid of. And we'll send each winner a nice, official-looking letter suitable for framing.
12. We might make some effort to find the entry with the most cows seen and the entry with the most breeds seen (entries with many sightings of rare breeds will be looked at suspiciously). These winners might get something too, maybe an autographed copy of this book.

A Field Guide to Cows Christmas Cow Count Officially Unofficial Entry Form

Date of the count

Place of the count	State:	County:
	Nearest town with a newspaper:	

Breed	Number seen

Submitted by	Witnessed by
Name:	Name:
Address:	Address:
Signature:	Signature:

(Submitted by and Witnessed by should be different people, if possible.)

Legal Cow Count Stuff

1. You are on your own out there. We are not responsible for anything that might happen to you, including but by no means limited to the following cow hazards: stepped-on feet, horn wounds, butted backsides, bite marks, bad smells, cow licks, hurt feelings, messed-up shoes, stalled or stuck cars, frostbite, sunburn, lactose intolerance, hay fever, boredom, confusion, frustration, or indifference.
2. Entries are welcome from everyone. But in the interest of fairness, professional cow watchers will not be eligible for prizes. Members of the National Bovine Appreciation Society's Cow Watchers Hall of Fame and Ice Cream Parlor, and their families, are also not eligible.